Universitext

Achim Bachem Walter Kern

Linear Programming Duality

An Introduction to Oriented Matroids

With 40 Figures

Springer-Verlag
Berlin Heidelberg New York
London Paris Tokyo
Hong Kong Barcelona
Budapest

Achim Bachem

Mathematisches Institut, Universität zu Köln,
Weyertal 86–90, W-5000 Köln 41, Fed. Rep. of Germany

Walter Kern

Department of Applied Mathematics, University of Twente,
P. O. Box 217, NL-7500 AE Enschede, The Netherlands

Mathematics Subject Classification (1991): 05B35, 90C05, 90C27, 90C48

ISBN 3-540-55417-3 Springer-Verlag Berlin Heidelberg New York
ISBN 0-387-55417-3 Springer-Verlag New York Berlin Heidelberg

Library of Congress Cataloging-in-Publication Data
Bachem, A. Linear programming duality: an introduction to oriented matroids / Achim Bachem, Walter Kern. p. cm. – (Universitext) Includes bibliographical references and index. ISBN 3-540-55417-3 (alk. paper). – ISBN 0-387-55417-3 (U.S.: alk. paper) 1. Matroids. 2. Linear programming. I. Kern, Walter, 1957– . II. Title. QA166.6.B33 1992 511'.6–dc20 92-14021

Printed in Germany

Typesetting: Camera ready by author
41/3140-543210 – Printed on acid-free paper

Preface

The main theorem of Linear Programming Duality, relating a "primal" Linear Programming problem to its "dual" and vice versa, can be seen as a statement about sign patterns of vectors in complementary subspaces of $\mathcal{R}^n$.

This observation, first made by R.T. Rockafellar in the late sixties, led to the introduction of certain systems of sign vectors, called "oriented matroids". Indeed, when oriented matroids came into being in the early seventies, one of the main issues was to study the fundamental principles underlying Linear Programming Duality in this abstract setting.

In the present book we tried to follow this approach, i.e., rather than starting out from ordinary (unoriented) matroid theory, we preferred to develop oriented matroids directly as appropriate abstractions of linear subspaces. Thus, the way we introduce oriented matroids makes clear that these structures are the most general - and hence, the most simple - ones in which Linear Programming Duality results can be stated and proved. We hope that this helps to get a better understanding of LP-Duality for those who have learned about it before und a good introduction for those who have not.

The present book arose from an early draft on polyhedral theory and two graduate courses held at Köln university in 1989/1990. No specific prerequisits are assumed. Basically, all we require is some familarity with linear algebra. The book is intended as an introduction to oriented matroid theory for the non-expert. Therefore we restricted ourselves to a thorough discussion of rather elementary results such as linear duality theorems and abstract polyhedral theory.

The only more advanced topic we treat is the topological realization of oriented matroids due to J.Folkman and J.Lawrence, resp. J.Edmonds and A.Mandel. A further reason for restricting ourselves in this way was the fact that at about the same time as this book will appear, a much more comprehensive treatment of oriented matroids by A.Björner, M.Las Vergnas, B.Sturmfels, N.White and G.Ziegler including an extensive discussion of more advanced topics will be available. Instead of repeating a reference to this work at the end of each chapter of our book we would like to recommend it here, once and for all, to the interested reader.

We are grateful to several colleagues for many stimulating discussions on the topic of this book. We are particularly grateful to Michael Hofmeister und Winfried Hochstättler who read the "beta version" of the text and provided extensive comments and suggestions. We have worked on this book and written the text at various universities. We acknowledge, in particular, the support of the German Research Association (DFG), the universities of Bonn, Köln and Twente.

Our special thanks are due to Frank Neuhaus for the efficient and careful typing the text in TEX.

Köln, October 1991

Achim Bachem
Walter Kern

Contents

Notation

We use standard notation throughout as far as possible and there are only very few things that are worth being mentioned explicitely:

Quantifiers are denoted by $\exists$ and $\forall$. "$\subseteq$" stands for (set theoretical) containment and "$\subset$" always means proper containment. Disjoint union will be denoted by "$\dot\cup$".

If A is an $m \times n$ matrix, then
$A_{i\cdot}$, $i \in \{1,\ldots,m\}$ denotes the i-th row of A and
$A_{\cdot j}$, $j \in \{1,\ldots,n\}$ denotes the j-th column of A.

More generally, if $I \subseteq \{1,\ldots,m\}$ and $J \subseteq \{1,\ldots,n\}$, then $A_{I\cdot}$ denotes the matrix made up by the rows indexed by I and $A_{\cdot J}$ denotes the matrix made up by the columns indexed by J. A_{IJ} denotes the submatrix of elements indexed by $(i,j) \in I \times J$. The transpose of a matrix A and a vector x are denoted by A^T and x^T, resp. Sometimes, transposition of vectors is omitted, in case no misunderstanding is possible. Thus, usually, vectors are implicitely assumed to be column vectors, but they may be used as row vectors sometimes without making the transposition explicitely. If A and B are subsets of a vectorspace, then $A+B$ denotes the set $\{a+b \mid a \in A,\ b \in B\}$. If Λ is a set of scalars, then $\Lambda A = \{\lambda a \mid \lambda \in \Lambda,\ a \in A\}$.

The set of all subsets of a set E is denoted by 2^E.
$\mathbb{Z}$ denotes the set of integers and $\mathbb{N}$ denotes the the set of positive integers.

Chapter 1

Prerequisites

This chapter introduces some notation and reviews basic facts from different fields. The reader is assumed to be familiar with the very elementary concepts in linear algebra and topology of Euclidean n-space as summarized in Section 1.2 and 1.3. Apart from these, the book is essentially selfcontained (with a minor exception of Chapter 9). Section 1.1 introduces the notion of partially ordered sets which is not treated in depth anywhere in this book, but is used to simplify the notation only. Section 1.4 introduces some notions from polyhedral theory.

1.1 Sets and Relations

We will use the standard notations from set theory throughout. Thus "$\in$", "$\cap$", "$\cup$", "$\subseteq$" etc. will have the usual meaning. We would like to point out, however, that "$\subset$" always stands for proper containment.

If X and Y are sets, then $R \subseteq X \times Y$ is called a **relation.** A **binary relation** on X is a relation $R \subseteq X \times X$. If $x, y \in X$, then we will, as usual, write xRy instead of $(x, y) \in R$.

Definition 1.1 *Let P be a set, and let $\leq$ be a binary relation on P such that*

(i) $p \leq p$ *for every* $p \in P$ *("reflexivity")*

(ii) $p \leq q$ *and* $q \leq p$ *implies* $p = q$ *("antisymmetry")*

(iii) $p \leq q$ *and* $q \leq r$ *implies* $p \leq r$ *("transitivity")*

Then $\leq$ is said to be a **partial order** *on P and the pair $(P, \leq)$ is called a* **partially ordered set,** *or briefly a* **poset.** *If there is no ambiguity as far as the order $\leq$ is concerned, we will also simply call P a poset. If any two elements $p, q \in P$ are "comparable", i.e. either*

$p \leq q$ or $q \leq p$, then $\leq$ is said to be a **total order** *on P, and P is said to be a* **totally ordered set**.

Example 1.2

1. *Let $P = \mathbb{N}$, the set of natural numbers, and let $n \leq m$ if n divides m. Then $(\mathbb{N}, \leq)$ is a poset.*
2. *Let $P = \mathbb{R}$, the set of real numbers, and let $\leq$ denote the usual ordering. Then $(\mathbb{R}, \leq)$ is a totally ordered set.*
3. *Let $P = 2^E$, the set of all subsets of a set E, and let $\subseteq$ denote the inclusion order on P. Then $(2^E, \subseteq)$ is a poset.*
4. *Let P denote a finite set of "jobs" to be performed on a machine, and let $p \leq q$ if $p = q$ or job p has to be executed before job q (say q uses some of the results that are computed during the execution of p). Then $(P, \leq)$ is a poset, provided there exists a "feasible schedule" for all the jobs. This is equivalent to saying that there is no "cycle" $p_1 < p_2 < \ldots < p_k < p_1$.*

Posets have enjoyed an increasing interest during the last years, in particular within the field of socalled "discrete mathematics", the reason being that — due to their generality — they often allow a common treatment of various special cases.

This section is not to be considered as an introduction into the theory of posets. (One would have to write a whole book on that subject just in order to cover the most fundamental developments.) All we want to do here is to introduce some notation and to present some simple facts about the existence of "dimension functions".

All posets P we will consider will be finite, and have a unique minimal element which is usually denoted by 0_P or simply 0. In case there is also a unique maximal element, this will be denoted by 1_P or simply 1.

If P is a poset and $Q \subseteq P$, then Q inherits an ordering from P in the obvious way. This is called the **induced order** on Q. If Q is such that $q \in Q$ and $x \leq q$ imply $x \in Q$ then Q is said to be **closed** or an **order ideal**. Obviously, the intersection of order ideals gives an order ideal again. Thus, if $A \subseteq P$, we may define the **closure** of A, denoted by $[A]$ to be the smallest order ideal containing A.

As usual, we will write "$<$" instead of "$\leq$ and $\neq$". If $p < q$ in a poset and there is no element "in between" p and q, i.e. $p \leq x \leq q$ implies either $x = p$ or $x = q$, then q is said to **cover** p and p is said to be **covered** by q. This is denoted by $p \lessdot q$.

A poset may be described in terms of its socalled **Hasse diagram** where each $p \in P$ is represented by a point and two points are linked by a straight line segment, in case one "covers" the other.

For example the poset $P = \{0, p_1, p_2, p_3, 1\}$, defined by the relations

$$\begin{aligned}&0 \leq 0\,,\, 0 \leq p_1\,,\, 0 \leq p_2\,,\, 0 \leq p_3\,,\, 0 \leq 1\,,\\ &p_1 \leq p_1\,,\, p_1 \leq p_3\,,\, p_1 \leq 1\,,\\ &p_2 \leq p_2\,,\, p_2 \leq p_3\,,\, p_2 \leq 1\,,\\ &p_3 \leq p_3\,,\, p_3 \leq 1\,,\\ &1 \leq 1\end{aligned}$$

has a Hasse diagram as indicated below

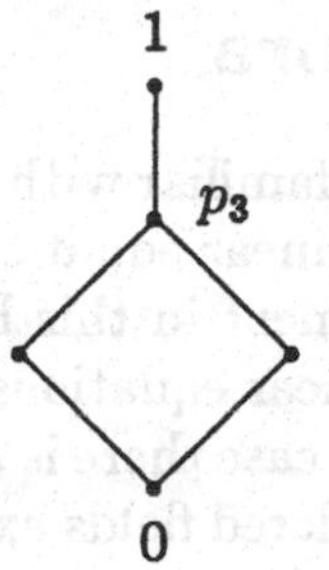

Figure 1.1: Hasse diagram

A **chain** between p and q in P is any sequence $p = p_0, p_1, \ldots, p_k = q$ such that $p_{i-1} < p_i$ for every i. The integer k is called the **length** of the chain. A **maximal chain** is one in which each p_i covers p_{i-1}.

P is said to have the JORDAN-DEDEKIND **chain property** if for any two elements $p, q \in P$ all maximal chains between p and q have the same length. We will also simply say that P is **JD** in that case.

Example 1.3 *Let $P = \{1, \ldots, n\}$ with $a \leq b$ if a divides b. Then P is JD.*

A **dimension function** on P is a function $\dim : P \to \mathbb{Z}$ such that $\dim 0 = -1$ and $\dim p = \dim q - 1$ whenever q covers p. One immediately concludes that a poset P has a dimension function if and only if it is JD. (The dimension function is obviously uniquely determined by $\dim 0 = -1$, so we are allowed to speak of *the* dimension function of P in this case.) In the example above, the dimension of an integer $k \in \{1, \ldots, n\}$ equals the number of its prime factors (counted with their multiplicities) minus 1. P is said to be **pure dimensional** if all its maximal elements have the same dimension d. This common dimension d is then said to be the **dimension** of P, denoted by dim P.

Let us finish this rather boring enumeration of definitions by introducing the following straightforward concept of isomorphism:

Two posets $(P_1, \leq_1)$ and $(P_2, \leq_2)$ are called **isomorphic** if there exists a bijection $\varphi : P_1 \rightarrow P_2$ such that

$$p \leq_1 q \quad \text{if and only if} \quad \varphi(p) \leq_2 \varphi(q).$$

$(P_1, \leq_1)$ and $(P_2, \leq_2)$ are called **antiisomorphic** if $(P_1, \leq_1)$ and $(P_2, \geq_2)$ are isomorphic.

1.2 Linear Algebra

The reader is assumed to be familiar with the basic concepts of linear algebra, such as systems of linear equations, subspaces and orthogonality. Much of the development in this book (if not all) is inspired by considering systems of linear equations and inequalities. Since inequalities make sense only in case there is an order relation $\leq$ around, we will have to deal with ordered fields exclusively.

Definition 1.4 *An* **ordered field** *F is a field endowed with a total ordering $\leq$ which is compatible with addition and multiplication, i.e. $\lambda \geq 0$ and $\mu \geq 0$ imply both $\lambda + \mu \geq 0$ and $\lambda \cdot \mu \geq 0$. If F is an ordered field, then $F_+ := \{\lambda \in F \mid \lambda \geq 0\}$ and $F_- := \{\lambda \in F \mid \lambda \leq 0\}$.*

The term "field" will always mean "commutative field". The two most important examples of such fields are $\mathbf{Q}$, the field of rationals and $\mathbf{R}$, the field of reals. Throughout this book, $\mathbf{K}$ denotes any field in between these two, i.e. $\mathbf{Q} \subseteq \mathbf{K} \subseteq \mathbf{R}$. (The reader familiar with TARSKI's theorem will observe that nothing interesting happens beyond $\mathbf{R}$ from our point of view.)

All linear spaces will be finite dimensional, i.e. isomorphic to $\mathbf{K}^n$ for some $n \geq 0$. (By convention $\mathbf{K}^0 = \{0\}$.) If L is a linear subspace of $\mathbf{K}^n$, we will usually stress this fact by writing $L \leq \mathbf{K}^n$ rather than just $L \subseteq \mathbf{K}^n$. Sometimes (for example in Chapter 2), it will be convenient to index the coordinates of $\mathbf{K}^n$ by an arbitrary finite set E, i.e. to replace $\mathbf{K}^n$ by $\mathbf{K}^E$ where E is a finite set. If $x \in \mathbf{K}^n$ (or $\mathbf{K}^E$), then the set of nonzero coordinates of x is called the **support** of x, denoted by $supp\, x$. If $L \leq \mathbb{K}^n$ is a subspace and $x \in L$ is a nonzero vector with minimal support, i.e. $\emptyset \subset supp\, y \subseteq supp\, x$ and $y \in L$ imply $supp\, y = supp\, x$, then x is called an **elementary** vector of L. The set of elementary vectors of L is denoted by $elem\, L$.

The basic results from linear algebra which are relevant in our context are presented in the following four subsections, entitled "Subspaces and Matrices", "Orthogonality", "Linear, Affine and Convex Hulls" and "Systems of Linear Equations".

Subspaces and Matrices

If $A = (a_{ij}) \in \mathbb{K}^{m \times n}$ is an $m \times n$ matrix, then there are two subspaces of $\mathbb{K}^n$ associated to A in a natural way:

$$\begin{aligned} ker\, A &:= \{x \in \mathbb{K}^n \mid Ax = 0\} \quad \text{and} \\ im\, A^T &:= \{x \in \mathbb{K}^n \mid u^T A = x^T \text{ for some } u \in \mathbb{K}^m\}. \end{aligned}$$

These are called the **kernel** of A and the **image** of A^T, resp. (Tranposition, as you will have noticed, will always be indicated by an upper case T. Vectors are usually considered to be column vectors, but sometimes we will be less strict in our notation in case no misunderstanding is possible.)

Theorem 1.5 *Every subspace $L \leq \mathbb{K}^n$ can be represented in both ways, i.e. there exist matrices A and B such that*

$$L = ker\, A \quad and \quad L = im\, B^T.$$

□

The notation we use when working with matrices is standard. In particular, if $A \in \mathbb{K}^{m \times n}$ is an $m \times n$ matrix, then

$A_{i\cdot}$ denotes the i-th row of A and

$A_{\cdot j}$ denotes the j-th column of A.

More generally, if $I \subseteq \{1, \ldots, m\}$ and $J \subseteq \{1, \ldots, n\}$, then $A_{I\cdot}$ denotes the submatrix made up of the rows $A_{i\cdot}, i \in I$, and $A_{\cdot J}$ denotes the submatrix made up of the columns $A_{\cdot j}, j \in J$. A_{IJ} denotes the submatrix whose elements are indexed by $(i, j) \in I \times J$.

Orthogonality

The **inner product** of two vectors $x, y \in \mathbb{K}^n$ is defined to be

$$x^T y = \sum_{i=1}^{n} x_i y_i \in \mathbb{K}$$

Two vectors $x, y \in \mathbb{K}^n$ are said to be **orthogonal**, denoted by $x \perp y$ if their inner product equals zero. The **orthogonal complement** of a set $S \subseteq \mathbb{K}^n$ is defined to be

$$S^\perp := \{y \in \mathbb{K}^n \mid x \perp y \text{ for all } x \in S\}.$$

Theorem 1.6

(i) *For any set $S \subseteq \mathbb{K}^n$, $S^\perp$ is a linear subspace of $\mathbb{K}^n$.*

(ii) *If $L \leq \mathbb{K}^n$ is a subspace, then $\mathbb{K}^n = L \oplus L^\perp$, i.e. every $x \in \mathbb{K}^n$ can be uniquely written as $x = u + v$ with $u \in L$ and $v \in L^\perp$. u and v are called the* **projections** *of x onto L and $L^\perp$, resp.*

(iii) *If $L = ker\, A$, then $L^\perp = im\, A^T$. If $L = im\, A$, then $L^\perp = ker\, A^T$.*

(iv) *$L^{\perp\perp} = L$ for every subspace $L \leq \mathbb{K}^n$.*

□

Linear, Affine and Convex Hulls

The **linear hull** of a set $S \subseteq \mathbb{K}^n$ is the smallest subspace $L \leq \mathbb{K}^n$ containing S. This is denoted by *lin* S. Alternatively, this may also be defined as

$$lin\, S = S^{\perp\perp}.$$

lin S may be explicitely described as the set of all linear combinations of elements of S, i.e.

$$lin\, S = \left\{ \sum_{i=1}^{k} \lambda_i x_i \mid x_i \in S, \lambda_i \in \mathbb{K}, k \in \mathbb{N} \right\}.$$

The **dimension** of a subspace $L \leq \mathbb{K}^n$, denoted by dim L, is defined to be the maximum cardinality of a linearly independent subset of L. Thus $\dim\{0\} = 0$. It is wellknown that $\dim\, L + \dim\, L^\perp = n$ for any subspace $L \leq \mathbb{K}^n$.

An **affine combination** of elements $x_1, \ldots, x_k \in \mathbb{K}^n$ is any expression of the form

$$\lambda_1 x_1 + \ldots + \lambda_k x_k \text{ with } \lambda_1, \ldots, \lambda_k \in \mathbb{K} \text{ and } \sum_i \lambda_i = 1$$

An **affine subspace** is a set $S \subseteq \mathbb{K}^n$ which is closed under taking affine combinations (of elements of S). The **affine hull** of a set $S \subseteq \mathbb{K}^n$ is the smallest affine subspace containing S. This is an affine subspace which may be explicitely described as the set of all affine combinations of elements of S.

Every nonempty affine subspace H can be written as

$$H = x + L$$

where x is an arbitrary element of H and L is a linear subspace of $\mathbf{K}^n$. The **dimension** of H is defined to be the dimension of L. The dimension of the empty set is defined to be -1.

A **convex combination** of elements $x_1, \ldots, x_k \in \mathbf{K}^n$ is any expression of the form

$$\lambda_1 x_1 + \ldots + \lambda_k x_k \quad \text{with } \lambda_1, \ldots, \lambda_k \in \mathbf{K}_+ \quad \text{and } \sum_i \lambda_i = 1.$$

A set $S \subseteq \mathbf{K}^n$ is **convex** if it is closed under taking convex combinations (of elements of S). The **convex hull** of a set $S \subseteq \mathbf{K}^n$, denoted by *conv* S is the smallest convex set containing S. This may be explicitely desribed as the set of all convex combinations of elements of S.

A convex set $S \subseteq \mathbf{K}^n$ is called a **cone** if it is nonempty and closed under nonnegative scalar multiplication, i.e. if $x \in S$ implies $\lambda x \in S$ for every $\lambda \in \mathbf{K}_+$. The **conic hull** of a set $S \subseteq \mathbf{K}^n$ is the smallest cone containing S. This is denoted by *cone* S.

Obviously, if $S \subseteq \mathbf{K}^n$ is a convex set, then

$$lin\, S = \{\lambda x - \mu y \mid \lambda, \mu \in \mathbf{K}_+, x, y \in S\}$$

and if $S \subseteq \mathbf{K}^n$ is a cone, then

$$lin\, S = \{x - y \mid x, y \in S\}.$$

The study of general convex sets and cones is outside the scope of linear algebra, and in fact it is outside the scope of this book. However, as it will turn out later, in case $S \subseteq \mathbf{K}^n$ is finite, the associated sets $P = conv\, S$ and $C = cone\, S$ can well be studied by means of linear algebra.

Systems of Linear Equations

A **system of linear equations** is given by

$$Ax = b$$

where $A \in \mathbf{K}^{n \times m}$ is an $n \times m$ matrix and $b \in \mathbf{K}^n$ is a vector. The system is called **homogeneous** in case $b = 0$ and **inhomogeneous** otherwise. The set of solutions of a homogeneous system is clearly the linear subspace $L = ker\, A \leq \mathbb{K}^m$. In case $b \neq 0$, the set of solutions is either empty or an **affine space**, i.e. a subset $S \subseteq \mathbb{K}^m$ that can be written as

$$S = x^0 + L = \{x^0 + x \mid x \in L\}.$$

Here, obviously, $L = ker\, A$ and x^0 is any particular solution of $Ax = b$.

A system of linear inequalities is given by

$$Ax \leq b$$

where $A \in \mathbb{K}^{m \times n}$ and $b \in \mathbb{K}^n$. The system is called **homogeneous** in case $b = 0$ and **inhomogeneous** otherwise. The set of solutions of a homogeneous system is obviously a cone, the set of solutions of an inhomogeneous system is obviously a convex set. Since systems of linear inequalities are usually not covered by textbooks on linear algebra (although we think they should), we do not assume any familiarity with such systems here. You will find a little bit more about this subject in Section 1.4 and much more later on.

1.3 Topology

This section is to provide some very elementary facts from socalled "general topology". Topological arguments will enter our discussion only in Chapter 9, so one may prefer to postpone this section until the very end.

The reader is assumed to be familiar with the most elementary concepts in general topology. Only a very few will be needed in fact. In addition to these, some elementary results of "PL-topology" will be needed in Chapter 9. These will be developed there (as far as we thought it were reasonable).

If X is a topological space and $Y \subseteq X$ then Y, endowed with the topology induced from X, is called a **subspace** of X. If Y is a subspace of X and Z is a subspace of Y, then Z is a subspace of X, i.e. the topology on Z that is induced from Y is the same as that induced from X. This fact (which follows immediately from the definition of induced or "relative" topology) will be referred to as the **transitivity of relative topology**.

If X is a topological space and $A \subseteq X$, then $cl\, A$ denotes the closure of A, $int\, A$ denotes the interior of A, i.e. the largest open subset of X which is contained in A, and ∂A denotes the boundary of A, i.e. $\partial A = cl\, A \setminus int\, A$. In case there is some ambiguity as to which space we are referring, we will write $cl_X A$, $int_X A$ and $\partial_X A$.

No "exotic" topological spaces will occur at any time in our book. Essentially, we will work within the Euclidean space $\mathbb{R}^n$ or its subspace $\mathbb{K}^n$ where $\mathbb{K} \subseteq \mathbb{R}$ is an arbitrary ordered field (cf. Section 1.2). In particular, the following elementary facts about (topological) subspaces of $\mathbb{R}^n$ will be needed:

Let $B_n := \{x \in \mathbb{R}^n \mid \|x\| \leq 1\}$, i.e. B_n denotes the n-dimensional unit ball in $\mathbb{R}^n$, and let $S^{n-1} := \partial B_n = \{x \in \mathbb{R}^n \mid \|x\| = 1\}$. (By convention, $B_0 = \{0\}$ and $S^{-1} = \emptyset$.) If $x \in S^{n-1}$, then $S^{n-1} \setminus \{x\}$ is homeomorphic to $\mathbb{R}^{n-1}$ (as can be shown by means of stereographic projection). If $B \subseteq \mathbb{R}^n$ and $h : B_n \to B$ is a homeomorphism, then $h(int\, B_n) = int\, B$ and $h(\partial B_n) = \partial B$. This is a consequence of the so-called **Domain Invariance Theorem** cf., e.g. [DUGUNDJI, Chapter XVII, Section 3].

1.4 Polyhedra

Consider a system of linear inequalities

$$Ax \leq b$$

for some $A \in \mathbb{K}^{m \times n}$ and $b \in \mathbb{K}^m$. The set of solutions $x \in \mathbb{K}^n$ is called a **polyhedron**, denoted by $P(A, b)$. Thus

$$P(A, b) = \{x \in \mathbb{K}^n \mid Ax \leq b\}.$$

Sometimes we will have to deal with special systems of linear inequalities, consisting of a set of equalities $Ax = b$ and a set of nonnegativity constraints $x \geq 0$. The corresponding polyhedron will be denoted by $P^=(A, b)$. Thus

$$P^=(A, b) = \{x \in \mathbb{K}^n \mid Ax = b, x \geq 0\}.$$

Note that this is in fact a polyhedron, since

$$\begin{array}{rcl} Ax & = & b \\ x & \geq & 0 \end{array} \quad \Leftrightarrow \quad \begin{array}{rcr} Ax & \leq & b \\ -Ax & \leq & -b \\ -Ix & \leq & 0 \end{array}$$

where I denotes the identity matrix.

Each of the inequalities $A_{i.}x \leq b_i$ determines a closed affine halfspace of $\mathbb{K}^n$. Thus, a polyhedron may be alternatively be defined as the intersection of a finite number of closed halfspaces. Geometrically, this may be sketched as Figure 1.2.

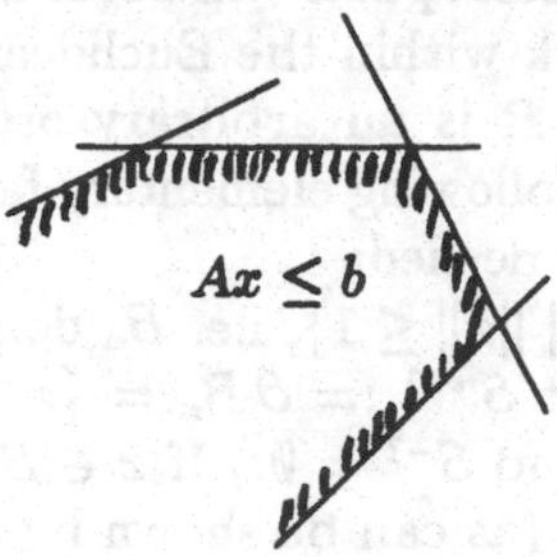

Figure 1.2: The polyhedron $P(A, b)$

A polyhedron $P \subseteq \mathbb{K}^n$ is said to be **bounded**, if P is a bounded subset of $\mathbb{K}^n$, i.e. if there exists $r \in \mathbb{K}$ such that $\|x\| \leq r$ for every $x \in P$. A bounded polyhedron will also be called a **polytope**.

The **recession cone** of a nonempty polyhedron $P \subseteq \mathbb{K}^n$ is defined to be the set

$$rec\,P := \{z \in \mathbb{K}^n \mid P + z \subseteq P\}.$$

The following lemma provides some alternative definitions of the recession cone:

Lemma 1.7 *Let $\emptyset \neq P = P(A, b) \subseteq \mathbb{K}^n$ be a polyhedron, and let $z \in \mathbb{K}^n$. Then the following are equivalent:*

(i) $z \in rec\,P$.

(ii) $\forall x \in P, x + \mathbb{K}_+ z \subseteq P$.

(iii) $\exists x \in P, x + \mathbb{K}_+ z \subseteq P$.

(iv) $Az \leq 0$.

Proof.

(i) ⇒ (ii): Let $z \in rec\,P$ and $x \in P$. By definition of $rec\,P$, this implies that $x_1 := x + z \in P$. This in turn implies that $x_2 := x_1 + z = x + 2z \in P$. Continuing this way, we inductively conclude that $x + \mathbb{N}z \subseteq P$. Since P is convex, this implies that $x + \mathbb{K}_+ z \subseteq P$.

(ii) ⇒ (iii): This is trivial (note that we assume $P \neq \emptyset$).

(iii) ⇒ (iv): Let $x \in P$ such that $x + \mathbb{K}_+ z \subseteq P$. By definition of $P = P(A, b)$, this means that $A(x + \lambda z) \leq b \quad \forall \lambda \in \mathbb{K}_+$, which implies $Az \leq 0$.

(iv) $\Rightarrow$ (i): Let $z \in \mathbf{K}^n$ such that $Az \leq 0$. Then $P + z \subseteq P$. In fact, if $x \in P$, i.e. $Ax \leq b$, then $A(x+z) \leq b + Az \leq b$, i.e. $x + z \in P$.

□

Example 1.8 *The following figure shows a polyhedron P and its recession cone $rec\,P$.*

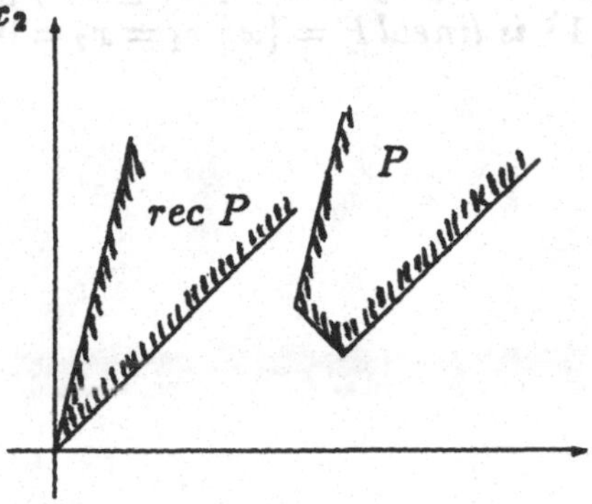

In general, a **polyhedral cone** is defined to be a polyhedron $C \subseteq \mathbf{K}^n$, which is a cone, i.e. $\mathbf{K}_+ C \subseteq C$. The following lemma gives an alternative definition:

Lemma 1.9 *Every polyhedral cone $C \subseteq \mathbf{K}^n$ can be described by a homogeneous system of inequalities, i.e. $C = P(A, 0)$. Conversely, any such set is a polyhedral cone.*

Proof. Let $C = P(A, b)$ be a polyhedral cone. We claim that C can be written as $C = P(A, 0)$. In fact, since C is a cone, we have $0 \in C$ and thus Lemma 1.7 implies that $0 + rec\,C = P(A, 0) \subseteq C$. On the other hand $C \subseteq rec\,C$ is trivial. Hence $C = rec\,C = P(A, 0)$. The converse, i.e. that every polyhedron $C = P(A, 0)$ is in fact a polyhedral cone, is obvious.

□

Linear subspaces $L \leq \mathbf{K}^n$ are special cases of polyhedral cones. In general, a polyhedral cone may or may not contain a linear subspace. Let us define for a general nonempty polyhedron $P \subseteq \mathbf{K}^n$ the set

$$lineal\,P := \{z \in \mathbf{K}^n \mid P + \mathbf{K}z \subseteq P\},$$

called the **lineality space** of P. This is obviously the largest subspace contained in $rec\,P$. Using Lemma 1.7 it is an easy exercise to show that for $P = P(A, b)$

$$\begin{aligned} lineal\,P &= \{z \in \mathbf{K}^n \mid z \in rec\,P \text{ and } -z \in rec\,P\} \\ &= \{z \in \mathbf{K}^n \mid Az = 0\} = ker\,A\,. \end{aligned}$$

Example 1.10 *The lineality space of $P \subseteq I\!\!K^3$, given by $P = \{x \in I\!\!K^3 \mid x_1 \geq 1, x_2 \geq 1\}$ is $lineal\,P = \{x \mid x_1 = x_2 = 0\}$.*

Chapter 2
Linear Duality in Graphs

Linear duality deals with the relationship between two complementary orthogonal subspaces L and $L^\perp$ of $\mathbf{K}^n$. The main theorem of linear duality, FARKAS' Lemma, will be presented in Chapter 4. In this chapter we will derive FARKAS' Lemma only for a special class of complementary pairs $(L, L^\perp)$ arising from directed graphs.

2.1 Some Definitions

Let V be a finite set, and let $E \subseteq V \times V$ be a finite family of ordered pairs $e = (u, v)$. Then $G = (V, E)$ is called a **directed graph** (or **digraph**, for short). The elements of V are called the **vertices** of G, and the elements of E are called the **edges** of G. Sometimes it is easier just to "draw" a graph rather than describing it explicitely by listing all its edges. For example, if $V = \{1, \ldots, 6\}$ and $E = \{(1,2), (1,5), (1,5), (1,6), (2,4), (2,4), (2,6), (3,2), (3,5), (4,5), (5,5), (5,5), (6,5)\}$, then $G = (V, E)$ may be drawn as indicated below (cf. figure 2.1).

Remark 2.1 *The term "directed" is due to the fact that edges are defined to be ordered pairs* $e = (u, v) \in V \times V$. *If one instead defines an edge to be an "unordered pair"* $e = \{u, v\}$ *with* $u, v \in V$ *(with* $u = v$ *not excluded), one arrives at the definition of* **graphs**. *We will always work with digraphs exclusively and therefore sometimes use the term "graph" as a synonym for "directed graph", since no misunderstanding is possible.*

This section is, of course, not considered to be an introduction into graph theory. The interested reader may consult one of the numerous textbooks on graph theory for that. In our context, graphs will only serve as a field where general (abstract) results from linear

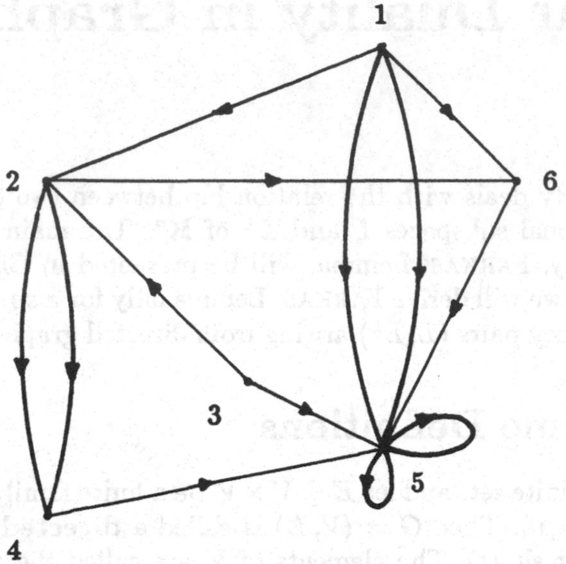

Figure 2.1

duality theory may be interpreted in a more intuitively appealing way. Therefore, we will try to get along with as few definitions as possible, introducing only those concepts which are necessary for developing linear duality theory in graphs.

Let $G = (V, E)$ be a graph. An edge $e = (u, v) \in E$ is said to **join** u to v, and u and v are called the **end vertices** of e. More precisely, v and u are called the **head and tail** of $e = (u, v)$, resp. Two vertices are also called **adjacent** if they are joined by an edge. We also say that u and v are **incident** with $e = (u, v)$ and vice versa. Two edges e and e' are called **incident** if they have at least one end vertex in common. In case e and e' have two end vertices in common, they are called **parallel** or **antiparallel**, according to whether the tail of e equals the tail of e' or the head of e'. An edge with just one end vertex is called a **loop**. A graph without loops and (anti-) parallel edges is called **simple**.

We say that $G' = (V', E')$ is a **subgraph** of $G = (V, E)$ if $V' \subseteq V$ and $E' \subseteq E$. If G' contains all edges of G whose end vertices are in V', then G' is said to be the **subgraph induced by** V', denoted by $G[V']$.

We shall often construct new graphs from old ones by "deleting" or "contracting" edges. If $e \in E$, then $G' := G\backslash e := (V, E\backslash e)$ is said to be obtained from G by **deleting the edge** e. More generally, if $A \subseteq E$, then $G \setminus A$ denotes the graph which is obtained from G by deleting all edges $e \in A$ (successively). If $e = (u, v) \in E$, then $G' := G/e$ denotes the graph which is obtained from G by identifying the vertices u and v and removing the edge e. We say that G/e is obtained from G by **contracting the edge** e. More generally, if $A \subseteq E$, then G/A denotes the graph obtained from G contracting all edges $e \in A$ (successively). It is easily seen that these two operations commute, i.e. if A and B are two disjoint subsets of E, then $G \setminus A/B = G/B \setminus A$. The graph $G \setminus A/B$ is called a **minor** of G.

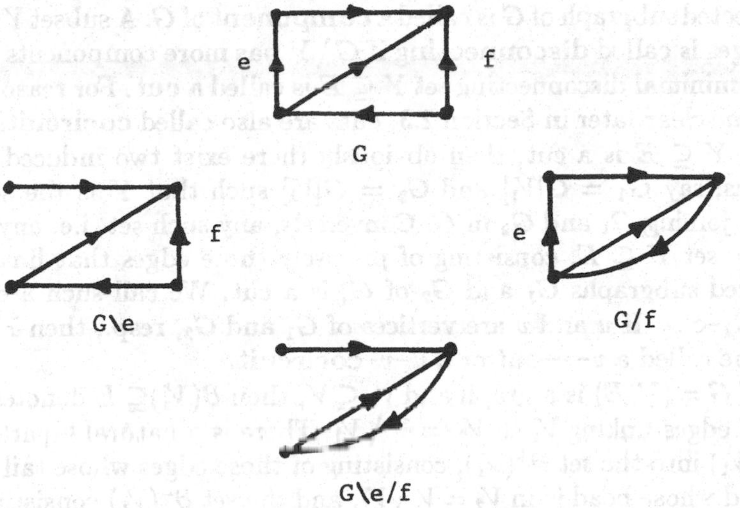

Figure 2.2: Contraction and Deletion

A **path** in G is an alternating sequence p of vertices and edges, say $p = (v_0, e_1, v_1, \ldots, e_k, v_k)$ such that each e_i has endvertices v_{i-1} and v_i, and the edges are pairwise distinct. (This latter restriction is usually not made in the definition of paths.) If $p = (v_0, e_1, v_1, \ldots, e_k, v_k)$ is a path, then v_0 and v_k are called its **initial** and **terminal vertex**, and k is called its **length**. We also say that p is a **path from** v_0 **to** v_k. The path p is called **simple** if the vertices $v_0, \ldots, v_{k-1}$ are pairwise distinct. The path p is called **closed** if its initial and terminal vertex coincide. The closed path p is a **circuit** if the subpath with initial vertex v_0 and terminal vertex v_{k-1} is simple.

If $p = (v_0, e_1, v_1, \ldots, e_k, v_k)$ is a path, then e_i is called a **forward** or **backward** edge, according to whether $e_i = (v_{i-1}, v_i)$ or (v_i, v_{i-1}). Usually, a path is not uniquely determined by its forward and backward edges but this minor ambiguity will not be essential in our context. Therefore, we often identify a path p with its set $P \subseteq E$ of edges, partitioned into P^+ and P^-, the sets of forward and backward edges, resp. Note that if p is a circuit, then P^+ and P^- are essentially (i.e., up to interchanging them) determined by the whole set $P \subseteq E$ of edges of p. In fact, there are only two possible ways (opposite directions) to go along a circuit.

A graph $G = (V, E)$ is called **connected** if any two of its vertices are joined by a path. Otherwise, G is called **disconnected**. A maximal connected subgraph of G is called a **component** of G. A subset $Y \subseteq E$ of edges is called **disconnecting** if $G \setminus Y$ has more components than G. A minimal disconnecting set $Y \subseteq E$ is called a **cut**. For reasons to become clear later in Section 2.3, cuts are also called **cocircuits**.

If $Y \subseteq E$ is a cut, then obviously there exist two induced subgraphs, say $G_1 = G[V_1]$ and $G_2 = G[V_2]$ such that Y is the set of edges joining G_1 and G_2 in G. Conversely, any such set, i.e. any non empty set $Y \subseteq E$, consisting of precisely those edges that link two induced subgraphs G_1 and G_2 of G, is a cut. We call such a cut a **G_1–G_2–cut**. If u and v are vertices of G_1 and G_2, resp., then Y will also be called a **u–v–cut** or a **u–v–cocircuit**.

If $G = (V, E)$ is a graph and $V_1 \subseteq V$, then $\partial(V_1) \subseteq E$ denotes the set of edges linking V_1 to $V_2 := V \setminus V_1$. There is a natural bipartition of $\partial(V_1)$ into the set $\partial^+(V_1)$, consisting of those edges whose tail is in V_1 and whose head is in $V_2 = V \setminus V_1$, and the set $\partial^-(V_1)$ consisting of those edges whose tail is in $V_2 = V \setminus V_1$ and whose head is in V_1. By our above consideration, a cocircuit in G may be respresented by a set $Y = \partial(V_1) \subseteq E$, partitioned into $Y^+ = \partial^+(V_1)$ and $Y^- = \partial^-(V_1)$. Again, Y^+ and Y^- are called the set of **forward** and **backward** edges of Y, resp. Note that, as in the case of circuits, if Y is a cocircuit, then Y^+ and Y^- are essentially (i.e., up to interchanging them) uniquely determined.

2.2 FARKAS' Lemma for Graphs

If you consider a given graph $G = (V, E)$ as a system of "one way streets", then a u–v–path P with $P^- = \emptyset$ represents a feasible way to go from u to v. Now suppose someone (say, your boss) hands you a graph, together with two prescribed vertices u and v, and asks you to find out whether such a path P from u to v with $P^- = \emptyset$ exists. He does not consider you to be a very trustworthy person, so he will not be content with getting just a "yes" or "no" answer, but he in addition wants you to prove that your answer is correct.

Obviously, in case such a path exists, you will eventually find it by searching the graph in any systematic way. (The most simple and stupid way would be to list all subsets $P \subseteq E$ and check them one by one.) Once you have found a path as required, you may hand it to your boss as a proof, showing that your "yes" answer is correct. In case you find that no such path exists, is there a simple way to convince your boss that your "no" answer is correct? In fact, there is: Suppose that during your examination of the graph, you have found a u–v–cut $Y \subseteq E$, dissecting G into components G_1 and G_2 such that G_1 contains u and G_2 contains v and all edges in between are directed from G_2 to G_1, i.e. Y is a u–v–cut with $Y^+ = \emptyset$. This gives evidence to the fact that no u–v–path P with $P^- = \emptyset$ can exist. The following theorem states that such a cut always exists, provided your "no" answer is correct, i.e. in case there is no path from u to v as required.

Theorem 2.2 (FARKAS' Lemma for Graphs) *Let $G = (V, E)$ be a connected graph, and let $u, v \in V$. Then either*

a) *$\exists$ u–v–path P with $P^- = \emptyset$.*

or

b) *$\exists$ u–v–cut Y with $Y^+ = \emptyset$.*

but not both .

Proof. The proof is constructive, i.e. we will explicitely describe an algorithm which either finds a path as in a) or a cut as in b). The algorithm proceeds by constructing the set $V_1 \subseteq V$ of all vertices that can be reached by a directed path starting at u. Initially, the set V_1 contains only u itself.

$V_1 := \{u\}$

WHILE $\partial^+(V_1) \neq \emptyset$ and $v \notin V_1$ DO

BEGIN

let $e = (s,t) \in \partial^+(V_1)$, set $V_1 := V_1 \cup \{t\}$, and $label(t) := s$

END.

Obviously, the algorithm stops after at most $|V|$ steps. If it ends up with $v \in V_1$, we know that there exist a directed path from u to v. This may be traced back from v by means of the labels. If the algorithm stops with $v \notin V_1$, then $\partial(V_1) = \partial^-(V_1)$ is a cut as in b). □

This simple result is the main theorem of linear duality in the context of graphs. Note that the general idea is to replace the statement "there exists no ..." by an equivalent statement "there exists ...". This general aspect of linear duality will be emphasized also in Chapter 3 where we investigate linear duality in the context of optimization. Let us conclude this section by restating the above theorem in a slightly different version.

Corollary 2.3 (FARKAS's Lemma) *Let $G = (V, E)$ be a graph. Then for every $e \in E$ either*

a) *$\exists$ circuit $X \subseteq E$ such that $e \in X$ and $X^- = \emptyset$.*
or
b) *$\exists$ cocircuit $Y \subseteq E$ such that $e \in Y$ and $Y^+ = \emptyset$.*

but not both .

Proof. Let $e = (v, u)$ and apply Theorem 2.2 to the graph $G \setminus e$. Obviously, a circuit containing e is made up of e and a path from u to v.

□

2.3 Subspaces Associated with Graphs

In this section we will associate two complementary subspaces L and $L^\perp$ to a given graph and see how the duality results from Section 2.2 may be restated in terms of L and $L^\perp$.

Definition 2.4 *Let $G = (V, E)$ be a graph. Then $\mathbb{K}^E$ is called the* **edge space** *of G. (Recall from Section 1.2 that we consider the elements of $\mathbb{K}^E$ as vectors with coordinates indexed by E, rather than functions from E to $\mathbb{K}$.) If $X \subseteq E$ is a circuit, partitioned into X^+ and X^- (the sets of forward and backward edges), then $x \in \mathbb{K}^E$, defined by*

$$x_e = \begin{cases} +1 & \text{if } e \in X^+ \\ -1 & \text{if } e \in X^- \\ 0 & \text{if } e \notin X \end{cases} \qquad (e \in E)$$

is called the **incidence vector** *of X.*

Similarily, if $Y \subseteq E$ is a cocircuit, partitioned into Y^+ and Y^-, then $y \in \mathbb{K}^E$, defined by

$$y_e = \begin{cases} +1 & \text{if } e \in Y^+ \\ -1 & \text{if } e \in Y^- \\ 0 & \text{if } e \notin Y \end{cases} \qquad (e \in E)$$

is called the **incidence vector** *of Y.*
(In the following, circuits and cocircuits will usually be denoted by upper case letters, $U, X, Y, Z \dots$ and their corresponding incidence vectors in $\mathbb{K}^E$ will be denoted by lower case letters $u, x, y, z \dots$.)

The **circuit space** *of G is the subspace of $\mathbb{K}^E$ which is generated by the incidence vectors of the circuits of G. Similarily, the* **cocircuit space** *of G is the subspace of $\mathbb{K}^E$ which is generated by the incidence vectors of the cocircuits of G.*

Example 2.5 *Consider the graph drawn below.*

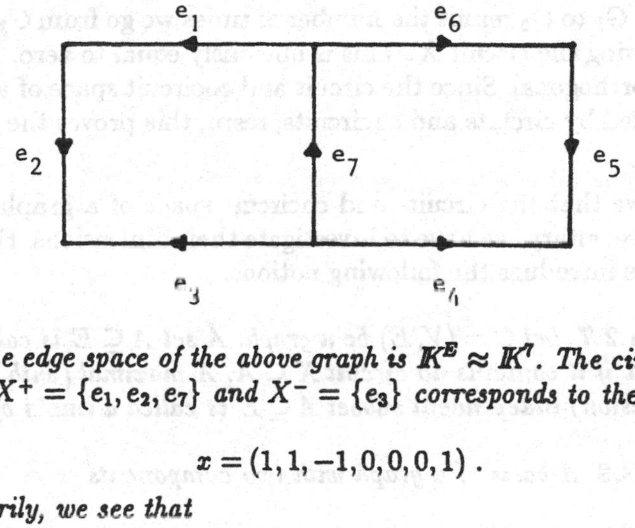

The edge space of the above graph is $\mathbb{K}^E \approx \mathbb{K}^7$. The circuit made up of $X^+ = \{e_1, e_2, e_7\}$ and $X^- = \{e_3\}$ corresponds to the incidence vector

$$x = (1, 1, -1, 0, 0, 0, 1)\ .$$

Similarily, we see that

$$x' = (0, 0, 0, -1, 1, 1, 1)$$

is the incidence vector of a circuit. Hence, for example,

$$x + x' = (1, 1, -1, -1, 1, 1, 2)$$

is an element of the circuit space. The two vectors

$$\begin{aligned} y &= (0, 0, 1, 0, -1, 0, 1) \qquad \text{and} \\ y' &= (-1, 0, -1, 0, 0, 0, 0) \end{aligned}$$

are incidence vectors of cocircuits. Hence, for example,

$$y + 3y' = (-1, 0, -2, 0, -1, 0, 1)$$

is an element of the cocircuit space.

In the following we are going to show that the circuit and cocircuit space of a graph form a complementary pair $L, L^{\perp}$ of subspaces in $\mathbb{K}^E$.

Lemma 2.6 *The circuit- and cocircuit space of a graph $G = (V, E)$ are orthogonal.*

Proof. Let $X \subseteq E$ be a circuit, partitioned into X^+ (its set of forward edges) and X^- (its set of backward edges). Let $Y \subseteq E$ be a cocircuit, dissecting G into G_1 and G_2 and assume that Y^+ is the set of edges directed from G_1 to G_2, and Y^- is the set of edges directed from G_2 to G_1. Let x and y, resp. denote the corresponding sign vectors. Now it is easily seen that the inner product $x^T y$ equals the number of times we go from G_1 to G_2 minus the number of times we go from G_2 to G_1, when following the circuit X. This is obviously equal to zero. Thus x and y are orthogonal. Since the circuit and cocircuit space of a graph are generated by circuits and cocircuits, resp., this proves the claim.

□

To prove that the circuit- and cocircuit space of a graph are in fact complementary, we have to investigate their dimensions. For that purpose, we introduce the following notions:

Definition 2.7 *Let $G = (V, E)$ be a graph. A set $A \subseteq E$ is called* **independent** *if it contains no circuit $X \subseteq A$. A maximal (with respect to set inclusion) independent subset $A \subseteq E$ is called a* **basis** *of G.*

Example 2.8 *A basis of a graph with two components*

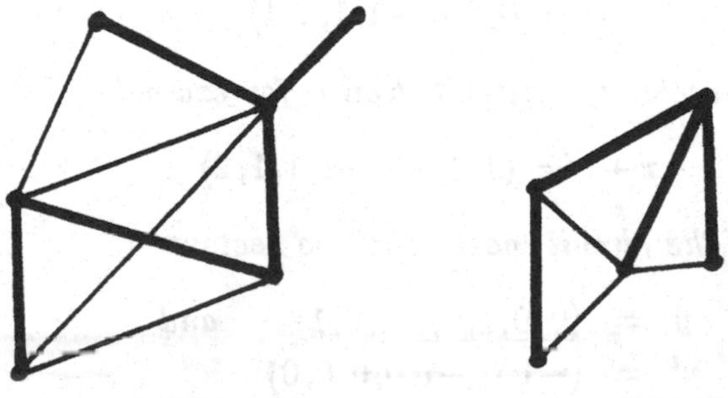

Lemma 2.9 *Let $G = (V, E)$ be a graph with n vertices and k components. Then every basis of G has precisely $n - k$ elements.*

Proof. The proof is by induction on n. If $n = k$, i.e. G consists of n isolated vertices, the claim is trivially true. Thus assume $n > k$, and let A be a basis of G. Choose any edge $e \in A$, and consider $G' := G/e$. Obviously, G' again consists of precisely k components. Furthermore, it is immediate that $A \setminus e$ is a basis of G'. In fact, every circuit in G' either corresponds to a circuit in G or to a path between the two end vertices of e. Hence, by induction, $|A \setminus e| = n - 1 - k$, which proves the claim.

□

Lemma 2.10 *Let $G = (V, E)$ be a graph with n vertices, m edges and k components. Then the dimension of its circuit space is at least $m - n + k$, and the dimension of its cocircuit space is at least $n - k$.*

Proof. Assume first that G is connected, i.e. $k = 1$. Let $A \subseteq E$ be a basis of G. Then any two vertices of G are linked by a unique path $P \subseteq A$. (In fact, such a path exists since (V, A) is connected, and it is unique since A does not contain a circuit.) For every $e \in E \setminus A$ let $X(e)$ denote the circuit of G that is made up of the unique path $P \subseteq A$ linking the two end vertices of e, and the edge e itself, say as a forward edge. Let $x(e) \in \mathbb{K}^E$ denote the corresponding incidence vectors. These form a set of $|E \setminus A| = m - n + 1$ vectors in the circuit space of G which is linearly independent. In fact, for each particular $e \in E \setminus A$, the vector $x(e)$ has a "+1" entry in coordinate e while all other vectors have coordinate e equal to 0. Thus the dimension of the circuit space is at least $m - n + 1$.

Next consider the cocircuits. If $e \in A$, then $G' = (V, A \setminus e)$ consists of two components, say G_1 and G_2. Let $Y(e)$ denote the cocircuit of G consisting of those edges joining G_1 and G_2 in G. These cocircuits give rise to a set of $|A| = n-1$ independent incidence vectors $y(e), e \in A$, in the cocircuit space of G. In fact, for a given $e \in A$, $y(e)$ has a nonzero coordinate e, while all the others have a zero entry in coordinate e. Thus the cocircuit space of G has dimension at least $n - 1$.

The proof is finished for $k = 1$. In case G has more components, the result follows easily by choosing independent circuits and cocircuits in each component.

□

The last two lemmata may be combined to obtain the following fundamental result:

Theorem 2.11 *The circuit- and cocircuit space of a graph $G = (V, E)$ form a complementary pair $L, L^{\perp} \leq \mathbb{K}^E$.*

□

The linear duality theorem we proved in Section 2.2 may thus be restated as follows:

Theorem 2.12 (Farkas' Lemma) *Let $L, L^{\perp} \leq \mathbb{K}^E$ denote the circuit- and cocircuit space of a graph $G = (V, E)$. Then for every $e \in E$ either*

a) $\exists\, x \in L\ , e \in supp\, x, x \geq 0$

or

b) $\exists\, y \in L^{\perp}, e \in supp\, y, y \leq 0$

but not both.

□

(Note that the "either – or" part follows from the corresponding version of these two theorems in Section 2.2. The appendix "but not both" is trivial since if x is as in a) and y as in b), then $x^T y < 0$, i.e., x and y are not orthogonal.)

The general principle behind is the main theorem of linear duality, stating that the above result remains valid if L is an arbitrary subspace of $\mathbb{K}^E$. This will be proved in Chapter 4. Maybe, the difference between subspaces associated to graphs in the above way and general subspaces of $\mathbb{K}^E$ is not very clear at this point. In the following we will represent the circuit- and cocircuit spaces of graphs in a way which makes their special structure more evident.

We know from linear algebra (cf. Section 1.2) that every complementary pair of subspaces $L, L^{\perp}$ in $\mathbb{K}^E$ can be represented by a matrix A such that $L = ker\, A$ and $L^{\perp} = im\, A^T$. In the case of "graphical" subspaces there is a canonical choice for the matrix A: Let $G = (V, E)$ with $V = \{v_1, \ldots, v_n\}$ and $E = \{e_1, \ldots, e_m\}$ without loops. Let $A = (a_{ij}) \in \mathbb{K}^{n \times m}$ be defined as

$$a_{ij} = \begin{cases} +1 & \text{if } v_i \text{ is the head of } e_j, \\ -1 & \text{if } v_i \text{ is the tail of } e_j, \\ 0 & \text{if } v_i \text{ and } e_j \text{ are not incident.} \end{cases}$$

This matrix A is called the **incidence matrix** of G. Thus each edge $e_j = (v_i, v_k)$ of G is represented by a column

$$A_{\cdot j} = \begin{pmatrix} 0 \\ \vdots \\ 0 \\ -1 \\ 0 \\ \vdots \\ 0 \\ 1 \\ 0 \\ \vdots \\ 0 \end{pmatrix} \begin{matrix} \\ \\ \\ \leftarrow i \\ \\ \\ \\ \leftarrow k \\ \\ \\ \\ \end{matrix}$$

One may also define incidence matrices for graphs containing loops by adding a zero column for each loop.

Each vertex v_i of G is represented by a row $A_{i\cdot}$ which has a "1" entry for every edge that has v_i as its head and a "−1" entry for every edge that has v_i as its tail.

Example 2.13 *Consider the graph $G = (V, E)$ as drawn below.*

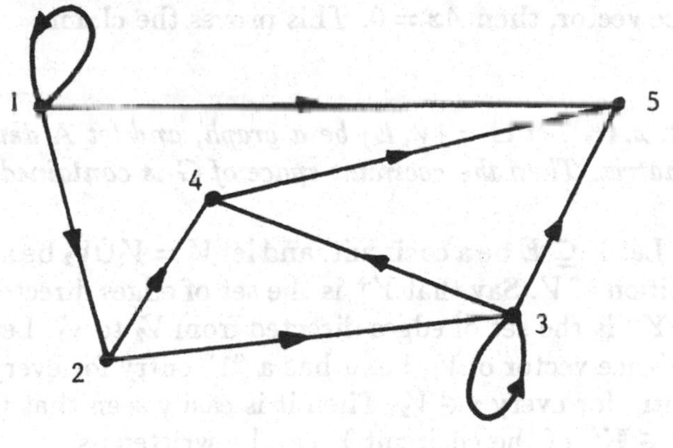

Its incidence matrix is given by

$$A = \begin{pmatrix} -1 & 0 & 0 & -1 & 0 & 0 & 0 & 0 & 0 \\ 1 & -1 & 0 & 0 & -1 & 0 & 0 & 0 & 0 \\ 0 & 1 & -1 & 0 & 0 & -1 & 0 & 0 & 0 \\ 0 & 0 & 0 & 0 & 1 & 1 & -1 & 0 & 0 \\ 0 & 0 & 1 & 1 & 0 & 0 & 1 & 0 & 0 \end{pmatrix}$$

Lemma 2.14 *Let $G = (V, E)$ be a graph, and let A denote its incidence matrix. Then the circuit space of G is contained in $\ker A$.*

Proof. Let $P = (v_0, e_1, v_1, \ldots, e_k, v_k)$ be a simple path. Consider its incidence vector $p \in \mathbf{K}^E$ which has a "+1" entry for each forward edge of P and a "−1" entry for each backward edge of P. Then it is obvious by induction on k that if $v_0 \neq v_k$, then

$$Ap = \begin{pmatrix} 0 \\ \vdots \\ 0 \\ -1 \\ 0 \\ \vdots \\ 0 \\ 1 \\ 0 \\ \vdots \\ 0 \end{pmatrix} \begin{matrix} \\ \\ \\ \leftarrow v_0 \\ \\ \\ \\ \leftarrow v_k \\ \\ \\ \\ \end{matrix}$$

From this it is immediate that if $X \subseteq E$ is a circuit (made up of a path from v_0 to v_k and an edge joining v_0 and v_k) and $x \in \mathbf{K}^E$ is its incidence vector, then $Ax = 0$. This proves the claim.

□

Lemma 2.15 *Let $G = (V, E)$ be a graph, and let A denote its incidence matrix. Then the cocircuit space of G is contained in $im\ A^T$.*

Proof. Let $Y \subseteq E$ be a cocircuit, and let $V = V_1 \dot{\cup} V_2$ be a corresponding partition of V. Say that Y^+ is the set of edges directed from V_1 to V_2, and Y^- is the set of edges directed from V_2 to V_1. Let $u \in \mathbf{K}^V$ be the incidence vector of V_1, i.e. u has a "1" entry for every $v \in V_1$ and a "0" entry for every $v \in V_2$. Then it is easily seen that the incidence vector $y \in \mathbf{K}^E$ of the cocircuit Y can be written as

$$y = u^T A,$$

i.e., $y \in im\ A^T$.

□

Theorem 2.16 *Let $G = (V, E)$ be a graph with incidence matrix A. Then the circuit- and cocircuit space of G are $\ker A$ and $im\ A^T$, resp..*

Proof. This is immediate from the two lemmata above and the fact that the circuit- and cocircuit space of a graph are complementary. □

This result tells us how to get the circuit- and cocircuit space of a given graph $G = (V, E)$. What about the converse problem, i.e. suppose we are given a subspace $L \leq \mathbb{K}^E$, say in terms of a matrix $\tilde{A}$ such that $L = ker\, \tilde{A}$, and we want to reconstruct the associated graph G (provided there is one)? The easiest way to do this, is to transform $\tilde{A}$ into an incidence matrix by means of elementary row operations (which do not affect the null space $ker\, \tilde{A}$). If one does not succeed, then L is not the circuit space of a graph. If one succeeds, the desired graph G is given by the incidence matrix A.

The following result tells us how the circuits and cocircuits of G may be obtained from the given circuit- and cocircuit spaces L, $L^\perp$ directly:

Proposition 2.17 *Let $G = (V, E)$ be a graph, and let $L, L^\perp \leq \mathbb{K}^E$ denote the circuit- and cocircuit spaces of G, resp. Then a vector $x \in \mathbb{K}^E$ is an elementary vector of L if and only if it is a scalar multiple of a circuit incidence vector of G. Similarily, a vector $y \in \mathbb{K}^E$ is an elementary vector of $L^\perp$ if and only if it is a scalar multiple of a cocircuit incidence vector of G.*

Proof. Let A denote the incidence matrix of G. We claim that $J \subseteq E$ is independent if and only if $A_{.J}$ consists of linearly independent columns. In fact, $A_{.J}$ is the incidence incidence matrix of $G' = (V, J)$. Hence $J \subseteq E$ is independent if and only if it contains no circuits, i.e. if and only if the circuit space of G', which is $ker\, A_{.J}$, is zero. This is equivalent to saying that the columns of $A_{.J}$ are linearly independent.

Now let $x \in \mathbb{K}^E$. If $x \in elem\, L$, then every proper subset $J \subset supp\, x$ is independent in G. But $X = supp\, x$ itself is dependent. This shows that X itself is a circuit of G. Since elementary vectors having the same support can differ only by a scalar multiple, the claim follows.

Next let us turn to the cocircuits. Let $z = u^T A$ be an elementary vector of the cocircuit space of G. Suppose that G consists of, say, k components, induced by $V_1, \ldots, V_k \subseteq V$. If u, considered as a map $V \to \mathbb{K}$, were constant on each component V_i, then obviously $z = 0$, i.e. z is not an elementary vector. Hence, for some V_i, there is a non-trivial partion $V_i = V_i' \dot\cup V_i''$ such that u is constant, say $u \equiv 1$, on V_i' and $u \neq 1$ on each vertex of V_i''. But this is obviously equivalent to saying that $supp\, z$ contains all edges joining V_i' to V_i''. Thus $supp\, z$ contains a cocircuit $Y \subseteq E$ and since z is elementary, this implies that $supp\, z = Y$. Hence every elementary vector in $im\, A^T$ is a scalar multiple of a cocircuit incidence vector. That cocircuit incidence vectors in turn are elementary vectors in $im\, A^T$ can be seen by essentially reversing the above argument. In fact, let $Y \subseteq E$ be a cocircuit, and let $y \in \mathbb{K}^E$ denote its incidence vector. If $V_i = V_i' \dot{\cup} V_i''$ is a partition of V_i such that Y consist of all edges joining V_i' and V_i'', then y can be written as $y = u^T A$ with $u \equiv 1$ on V_i', $u \equiv 0$ on V_i'' and u taking some constant value on each other V_j. If $z = w^T A$ is any nonzero vector with $supp\, z \subseteq supp\, y$ then w must also take constant values on each V_j, $j \neq i$. Furthermore, w must take a single constant value, say λ' on V_i' and a single constant value $\lambda'' \neq \lambda'$ on V_i''. But this implies already $supp\, z \supseteq supp\, y$. Hence y is in fact elementary.

□

2.4 Planar Graphs

This section may be considered as an appendix to Chapter 2 and therefore we will ommit some of the proofs. Our main object is to introduce a special class of graphs ("planar graphs"), where duality between circuits and cocircuits becomes most evident (or "visible").

A graph $G = (V, E)$ is called a **planar** graph if it can be drawn in the plane in such a way that no two edges cross each other. To put it a little bit more precisely, one may say that the edges are allowed to be simple JORDAN curves connecting the points corresponding to the end vertices of an edge and that by "non crossing" we mean that two edges connecting different vertices have no point in common except possibly one of their endpoints. One can show that this is equivalent to requiring every edge to be represented by a straight line segment, one for each class of (anti-) parallel edges and disregarding loops. Such a representation of G will be called a **plane** graph. There is a wellknown theorem, due to KURATOWSKI which characterizes planar graphs. This states that a graph $G = (V, E)$ is planar if and only if G has no minor "isomorphic" to one of the following two graphs

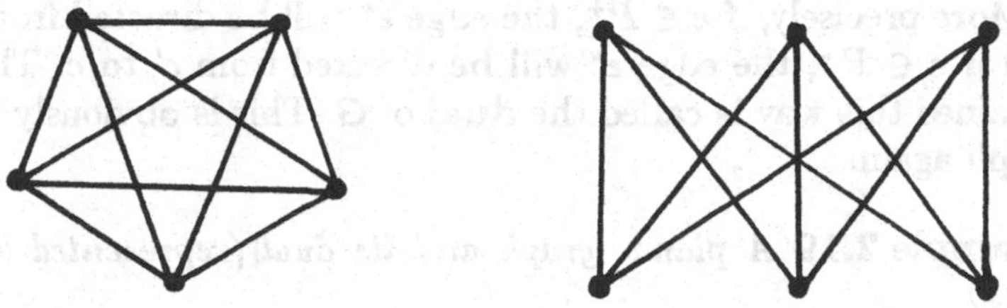

Known proofs of this result are rather long (though elementary) and therefore ommitted here.

Now consider a plane graph $G = (V, E)$. Removing its vertices and edges from the plane leaves a number of connected components, called **faces** or **countries**. Clearly each plane graph has exactly one unbounded face. The **boundary** of a face is the set of edges in its closure. Two countries are called **neighbouring** if their boundaries have an edge in common.

Let us mention the following result, known as EULER's **formula**, which we shall meet again in a more general context in Chapter 9.

Theorem 2.18 (EULER's **formula**) *Let $G = (V, E)$ be a connected plane graph with n vertices, m edges and f faces. Then*

$$n - m + f = 2.$$

Proof. We apply induction on the number of faces. If $f = 1$, then G has no circuits, i.e. E is a basis of G and hence $m = n - 1$ so the result holds. Now suppose that $f > 1$. Then there is a bounded face whose boundary obviously is a circuit $X \subseteq E$ of G. Hence deleting an edge $e \in E$ leaves the graph connected and decreases both f and m by 1. Thus the formula follows by induction.

□

The most important thing about planar graphs in our context is the concept of dual graphs. This is as follows. Suppose $G = (V, E)$ is a plane graph. Then we may define a new graph by considering each country as a vertex and by joining two vertices by an edge, provided the corresponding countries are neighbouring. More precisely, assume c is a bounded country, and let $P \subseteq E$ be the circuit consisting of the boundary of c such that P^+ and P^- are the forward and backward edges if we go around P in a counterclockwise orientation. Then for every edge $e \in P$ we include an edge e^* in the dual graph joining the country c to the neighbouring country c' which is separated from c by e. More precisely, if $e \in P^+$, the edge e^* will be directed from c to c' and if $e \in P^-$, the edge e^* will be directed from c' to c. The graph obtained this way is called the dual of G. This is obviously a planar graph again.

Example 2.19 *A planar graph and its dual (represented by dotted lines)*

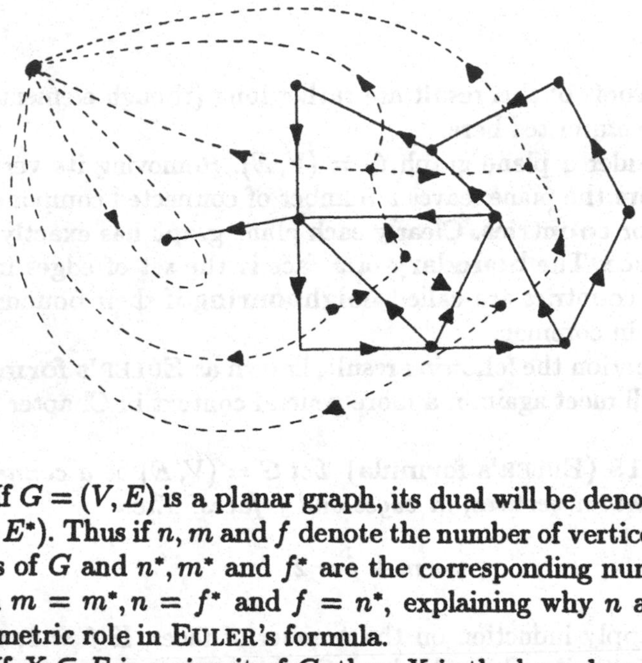

If $G = (V, E)$ is a planar graph, its dual will be denoted by $G^\perp = (V^*, E^*)$. Thus if n, m and f denote the number of vertices, edges and faces of G and n^*, m^* and f^* are the corresponding numbers of $G^\perp$, then $m = m^*, n = f^*$ and $f = n^*$, explaining why n and f play a symmetric role in EULER's formula.

If $X \subseteq E$ is a circuit of G, then X is the boundary of the union of faces it encludes. The corresponding set $X^* \subseteq E^*$ consists of those edges in E^* that join a country "within" X to a country "outside" X. This shows that every circuit of G corresponds to a cocircuit of $G^\perp$. (The details are left to the reader.) If G is connected, its circuit space has dimension $m - n + 1 = f - 1 = n^* - 1$, which is the dimension of the cocircuit space of $G^\perp$. Hence, by the observation above, the circuit space of G is equal to the cocircuit space $G^\perp$. (More precisely, the two spaces are isomorphic, the isomorphism being induced by the obvious map $\mathbf{K}^E \to \mathbf{K}^{E^*}$.) This further implies that the cocircuit space of G is the circuitspace of $G^\perp$. Since circuits and cocircuits can be identified as elementary vectors within the spaces they generate, this gives a one-to-one-correspondence between circuits of G and cocircuits of $G^\perp$.

Theorem 2.20 *Let $G = (V, E)$ be a connected planar graph, and let $G^* = (V^*, E^*)$. Then $X \subseteq E$ is a circuit of G if and only if $X^* \subseteq E^*$ is a cocircuit of G^*, and $Y \subseteq E$ is a cocircuit of G if and only if $Y^* \subseteq E^*$ is a circuit of G^*.*

□

2.5 Further Reading

Classical textbooks on graph theory are:

C. BERGE *The Theory of Graphs and its Application*, J. Wiley, New York (1962).

F. HARARY *Graph Theory*, Addison-Wesley, Reading, Mass. (1969).

Those who are interested in the history of graphs, may consult the book "Graph Theory 1736–1936" by N.L. BIGGS, E.K. LLOYD and R.J. WILSON, Oxford University Press, London (1976). Since 1960, the field has seen a real explosion of research and applications. We therefore list just a very few out of the huge number of more recent textbooks on graph theory:

B. BOLLOBÁS *Graph Theory*, Springer Heidelberg (1979).

C. BERGE *Graphs*, North-Holland, Amsterdam (1985).

M. GOUDRAN, M. MINOUX and S. VAJDA *Graphs and Algorithms*, J. Wiley, New York (1984).

W. T. TUTTE *Graph Theory, Encyclopedia of Mathematics and its applications, Vol 21*, Addison Wesley, Reading, Mass. (1984).

Chapter 3
Linear Duality and Optimization

3.1 Optimization Problems

Many problems in the real world (and in mathematics) can be formulated in the following way:

> Let $f : \mathbb{K}^n \rightarrow \mathbb{K}$ be a function, and let $S \subseteq \mathbb{K}^n$.
>
> Find $\bar{x} \in S$ such that $f(\bar{x}) \geq f(x)$ for all $x \in S$.

Such a problem is called a **mathematical programming problem (mathematical optimization problem)** or simply a **mathematical program.** More precisely, one may call it a **maximization problem**, because we are looking for an $\bar{x} \in S$ that maximizes the function f. Analogously, in case we seek for an $\bar{x} \in S$ with $f(\bar{x}) \leq f(x)$ for all $x \in S$ we speak of a **minimization problem.** The set $S \subseteq \mathbb{K}^n$ is called the **constraint set** or the **set of feasible solutions.** The function $f : \mathbb{K}^n \rightarrow \mathbb{K}$ is called the **objective function.** Any $\bar{x} \in S$ that maximizes (resp. minimizes) f on S is called an **optimal solution.** It is common to shorten the notation of the statement of an optimization problem by writing

$$\max_{x \in S} f(x)$$

although this is a slight misuse of the max operator in general. (In fact, it would be more appropriate to replace it by "sup", since optimal solutions may not exist in general, as discussed below.)

From the above statement of an optimization problem it is not quite clear what it means to "solve" an optimization problem. Let us examine this question in more detail. Note that optimal solutions may fail to exist for several reasons. First, it may happen that $S = \emptyset$, i.e. there are no feasible solutions at all. In this case the problem is called **inconsistent**. Otherwise, i.e. if $S \neq \emptyset$, the problem is called **consistent**. Whether or not a given problem is consistent may not be obvious in advance. (As discussed below, it may be arbitrarily hard to find out whether or not $S = \emptyset$, depending on how S is defined.) Secondly, it may happen that the objective function is **unbounded** on S, i.e. there exists a sequence $x_1, x_2, \dots$ of feasible solutions $x_i \in S$ such that $f(x_i) \geq i$ for every $i \in \mathbb{N}$. Obviously, in this case again no optimal solution exists. Finally, even in case the problem is **bounded**, i.e. there exists an upper bound $u \in \mathbb{R}$ such that $f(x) \leq u$ for every $x \in S$, it may happen that for every $x \in S$ there exists a $y \in S$ such that $f(x) < f(y)$. (For example, choose $S = [0, \pi] \cap \mathbb{Q}$, and let $f(x) := x$.) In this case the problem is called **degenerate**. Summarizing, a necessary condition for the existence of optimal solutions is that the problem is consistent, bounded and nondegenerate. **Solving** an optimization problem now amounts to find out and prove whether the problem is consistent/bounded/degenerate and in case optimal solutions exist, to find one and prove that it is indeed optimal.

Solving an optimization problem may be arbitrarily hard, depending on how complicated the constraint set S or the objective function f are allowed to be. In particular, there are classes of optimization problems that can provably not be solved by **any** algorithm. (This statement, of course does not mean anything, as long as one does not define precisely what should be understood by the term "algorithm". We do not want to give a rigorous definition here because this subject is outside the scope of our book. The interested reader is asked to consult any book on computability for more about this topic.) The main reason for the unsolvability of general optimization problems is GÖDEL's **incompleteness theorem** and the work arising from HILBERT's **tenth problem**, cf. the references given in Section 3.3. Consider for example the following class of problems. For each polynomial $p(x) = p(x_1, \dots, x_n)$ with integer coefficients in n variables $x_1, \dots, x_n$ let $f(x) := |p(x)|$ be the corresponding objective function, and let $S := \mathbb{Z}^n$ be the constraint set. Then solving the optimization problem

$$\begin{gathered} \min f(x) \\ x \in S \end{gathered}$$

is at least as difficult as deciding wether or not there exists an $x \in \mathbb{Z}^n$ such that $p(x) = 0$. One can prove that there is no algorithm which, given an arbitrary polynomial p as input, decides whether or not $p(x) = 0$ has a solution $x \in \mathbb{Z}^n$.

Thus one has to impose some restrictions on the set $S \subseteq \mathbb{K}^n$ and the function $f : \mathbb{K}^n \to \mathbb{K}$ in order to get classes of solvable problems. One possible restriction, for example, is to choose the constraint set as a convex set $S \subseteq \mathbb{K}^n$ and to allow only concave functions $f : \mathbb{K}^n \to \mathbb{K}$ (in case of a maximization problem). The class of problems arising this way is the class of socalled **convex optimization problems.** The interested reader is invited to consult any book on mathematical programming in order to learn about how to solve such kinds of problems or other restricted classes of optimization problems. Here we will be concerned only with a still more restrictive class of problems, the socalled **linear optimization problems**, which arise by taking $f : \mathbb{K}^n \to \mathbb{K}$ as a linear function, i.e.,

$$f(x) = c^T x$$

for some $c \in \mathbb{K}^n$, and by choosing the constraint set $S \subseteq \mathbb{K}^n$ to be a polyhedron

$$S = P(A, b) = \{x \in \mathbb{K}^n \mid Ax \leq b\}$$

for some $A \in \mathbb{K}^{m \times n}$ and $b \in \mathbb{K}^m$. Such a problem will usually be denoted by

$$\begin{array}{ccc} \max cx & \text{or} & \min cx \\ Ax \leq b & & Ax \leq b. \end{array}$$

(In the special case of linear optimization problems the use of the max operator will turn out to be justified. In fact, as we will see later, a linear optimization problem which is both consistent and bounded also has optimal solutions, i.e. there is no degenerate linear optimization problem.)

By definition, the constraint set of a linear optimization problem (or **linear programming problem** or a **linear program**, simply denoted by **LP**) is the intersection of a finite number of closed halfspaces in $\mathbb{K}^n$. Thus, geometrically, an LP in two dimensious (i.e. $n = 2$) may be sketched as below. Obviously, in the example indicated below, there is a unique optimal solution $\bar{x}$.

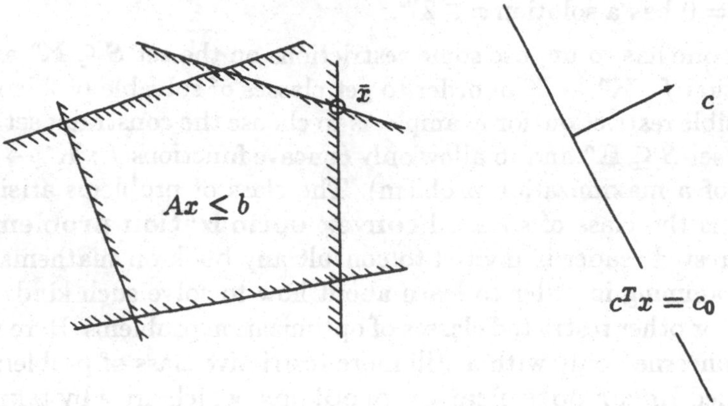

3.2 Recognizing Optimal Solutions

As mentioned already in Section 3.1, the class of linear programming problems is a rather restricted class of optimization problems. Yet there are interesting problems that may be stated as linear programs. Let us look at some examples just in order to get a feeling for what kind of problems may be formulated this way.

Example 3.1 (Assignment Problem) *Suppose we are given n machines and n people to work with these machines. Suppose in addition that man i when operating with machine j has a certain "skill", say $c_{ij} \geq 0$. Now we are to design a workplan for the next year, i.e. we are to find an assignment that maximizes the overall skill. In order to do so, we introduce variables x_{ij} indicating the amount of time (fraction of one year) that man i is to work with machine j. Our problem can then be stated as an* LP *in the following way.*

$$\begin{aligned} \max f(x) &= \textstyle\sum_{i,j} c_{ij} x_{ij} \\ \textit{subject to} \quad & \textstyle\sum_j x_{ij} \leq 1 \quad \forall i = 1, \ldots, n \\ & \textstyle\sum_i x_{ij} \leq 1 \quad \forall j = 1, \ldots, n \\ & x_{ij} \geq 0 \quad \forall i, j \end{aligned}$$

The objective function obviously gives the "overall skill". The inequalities describing the set of feasible solutions express that every man should work at most one year in total and that every machine should be used at most one year in total.

Example 3.2 (The Diet Problem) *One of the first problems that have ever been formulated as an* LP *is the following minimization problem: Consider a homemaker buying food. He has, say, a choice between n different foods, containing m nutrients each. Let*

a_{ij} := *amount of the i-th nutrient in one unit of the j-th food,*

r_i := *amount of the i-th nutrient he needs (in units per month),*

x_j := *his consumption of the j-th food (in units per month),*

c_j := *cost per unit of the j-th food.*

If he wants to find the least expensive diet satisfying his demands r_i, then he has to solve the following LP*:*

$$\min c^T x$$
$$Ax \geq r$$
$$x \geq 0.$$

Now suppose we want to write a computer program for solving an LP $\max\{c^T x \mid Ax \leq b\}$. Suppose that our problem data, i.e., A, b and c, are rationals, i.e. suppose that $\mathbb{K} = \mathbb{Q}$ for a moment. (Since a computer has only a finite number of symbols at its disposal, rational numbers are all we can represent in a straightforward way using a finite amount of storage only.) The simplest and most stupid computer program one can possibly think of is to do **exhaustive search**: Choose any enumeration $x_1, x_2, x_3, \ldots$ of $\mathbb{Q}^n$ and try successively whether x_i is feasible, i.e. whether $Ax_i \leq b$ or not, for $i = 1, 2, 3, \ldots$. Each time we encounter a feasible solution x_i, we compare its objective function value $c^T x_i$ with the largest one we have found so far, and proceed.

Besides from being ridiculously inefficient, this method suffers from a much more fundamental deficiency: Since we try all possible values $x \in \mathbb{Q}^n$ successively, we can be sure that we will eventually find an optimal solution, provided there is any. However, we won't recognize that, i.e. our program will run forever, even if it had found already an optimal solution.

Thus, what is missing is a stopping criterion which tells us whether our current best solution $x = x_i$ is optimal or not, i.e. which one of the following alternatives a) and b) holds:

a) $\exists y \in \mathbb{Q}^n$ s.t. $Ay \leq b$ and $c^T y > c^T x$.

b) $\forall y \in \mathbb{Q}^n$: $(Ay \leq b \Rightarrow c^T y \leq c^T x)$.

Note the similarity with the problem discussed in Section 2.2, where we had to convince our boss of the existence resp. nonexistence of certain paths in a graph. Proving the existence of such a path had been an easy thing while proving the nonexistence had been more difficult (although in principal no problem occurs in that case because the number of paths to be examined is still finite). Here, the situation is the very same. In order to prove a) one simply has to present a solution y with $c^T y > c^T x$ (and such a solution can be found by running our exhaustive search program above). However, proving that b) holds appears to be more difficult. This is precisely where linear duality comes in. As we mentioned already in Chapter 2, the main idea is to replace the statement b) by an equivalent statement b'), thereby replacing the all quantifier "$\forall$" by an existential quantifier "$\exists$". (This is much in the spirit of Section 2.2, where we have seen that the nonexistence of certain paths is equivalent to the existence of certain cuts.)

In the following we will show that proving b) essentially amounts to proving the emptiness of a polyhedron $P \subseteq \mathbb{K}^n$, given by a system of linear inequalities. Consider any current solution x of $Ax \leq b$. Let $I := \{i \mid A_{i.}x = b_i\}$. This is called the set of **active constraints** for x or the **equality set** of x. The polyhedral cone $C = P(A_{I.}, 0) = \{z \mid A_{I.}z \leq 0\}$ is called the **cone of feasible directions** for x and every $z \in C$ is called a **feasible direction** for x.

Lemma 3.3 *Let $P = P(A, b) \subseteq \mathbb{K}^n$ be a polyhedron, and let $x \in P$. Then $z \in \mathbb{K}^n$ is a feasible direction for x if and only if $y = x + \lambda z \in P$ for sufficiently small $\lambda \in \mathbb{K}, \lambda > 0$.*

Proof. This is a straightforward exercise. Let I denote the set of active constraints for x, and let $J := \{1, \ldots, m\} \setminus I$. Then $A_{I.}x = b_I$ and $A_{J.}x < b_J$. Thus if $z \in \mathbb{K}^n$, we may choose $\lambda > 0$ small enough to ensure that $A_{J.}(x + \lambda z) \leq b_J$. Thus $y = x + \lambda z \in P$ for sufficiently small $\lambda > 0$ if and only if $A_{I.}y \leq b_I$, i.e. $A_{I.}z \leq 0$, which means that z is a feasible direction for x.

□

Corollary 3.4 *Let $P = P(A, b) \subseteq \mathbb{K}^n$, let $x \in P$, and let I denote the set of active constraints for x. Let $Q := \{y \mid A_{I.}z \leq 0, c^T z = 1\}$. Then the two alternatives a) and b) from above are respectively equivalent to*

- **a')** *Q is nonempty.*
- **b')** *Q is empty.*

Proof. This is immediate from Lemma 3.3. Q is nonempty if and only if there is a feasible direction z satisfying $c^T z = 1$. This in turn is equivalent to saying that for $\lambda > 0$ sufficiently small there exists a better solution $y = x + \lambda z \in P$.

□

The linear duality theorem, which will be stated and proved in Chapter 4, says that to every polyhedron $Q \subseteq \mathbb{K}^n$ one can efficiently compute a polyhedron $Q' \subseteq \mathbb{K}^m$ such that

$$Q \text{ is empty} \Leftrightarrow Q' \text{ is non empty.}$$

This shows that (at least in principle) it is possible to decide which of the above alternatives a') and b') holds: We simply choose any enumeration $x_1, x_2, x_3 \ldots$ of the rational vectors $x \in \mathbb{Q}^n$ and successively check whether $x_i \in Q$ or $x_i \in Q'$. In any case, after a finite number of steps, we will eventually discover that Q is nonempty or Q' is nonempty.

3.3 Further Reading

The two most important historical papers on decidability are:

K. GÖDEL *On Formally Undecidable Propositions of Principia Mathematica and Related Systems I*, Monatshefte für Mathematik und Physik **38** (1931), pp. 173–198, reprinted in S. Ferman et al, "Kurt Gödel: Collected Works I: Publications 1929 – 1936", Oxford University Press (1986).

A.M. TURING *On Computable Numbers with an Application to the Entscheidungsproblem*, Proceedings London Mathematical Society **42** (1937), pp. 230–265, reprinted in M. Davis, "The Undecidable-Basic Papers on Undecidable Propositions, Unsolvable Problems and Computable Functions", Hewlett: Raven Press (1965).

Known proofs of the undecidability of HILBERT's Tenth Problem are rather involved. The interested reader may consult

J.P. JONES and MATIJASEVIC *Register Machine Proof of the Theorem on Exponential Diophantine Representation of Enumerable Sets*, Journal on Symbolic Logic **49** (1984), pp. 818–829,

presenting a simple proof of a related result and more references on that subject.

The following list contains some references on (general) optimization and linear optimization:

V. CHVÁTAL *Linear Programming*, W.M. Freeman, New York (1983).

G. B. DANTZIG *Linear Programming and Extensions*, Princeton University Press, Princeton, N. J. (1963).

R. FLETCHER *Optimization*, Academic Press (1969).

D. LUENBERGER *Introduction to Linear and Nonlinear Programming*, Addison Wesley (1973).

G. MCCORMICK *Nonlinear Programming*, John Wiley (1983).

K. MURTY *Linear and Combinatorial Programming*, John Wiley (1976).

A. SCHRIJVER *Theory of Linear and Integer Programmning* , John Wiley (1986).

A brief sketch of the history of linear programming is given by the "founder" of linear programming, George Dantzig, in his article "Reminiscenses about the Origin of Linear Programming", in:

A. BACHEM, M. GRÖTSCHEL and B. KORTE *Mathematical Programming. The State of the Art*, Springer (1983).

Chapter 4
The FARKAS Lemma

The result mentioned at the end of Section 3.2, which relates the emptiness of a given polyhedron to the nonemptiness of a certain other polyhedron is wellknown as the FARKAS Lemma. We will state this theorem in a more precise form in this section. Moreover, we shall give several equivalent formulations of the FARKAS Lemma, which we derive from each other by introducing standard techniques in polyhedral theory. In particular, we will show that the FARKAS Lemma can be seen as a theorem relating a subspace L of $\mathbf{K}^n$ to its orthogonal complement $L^\perp$ in the sense of Theorem 2.10.

4.1 A first version

Recall that our problem in Section 3.2 was to verify the emptiness of a polyhedron given by a set of linear inequalities.

Suppose $P = P(A, b)$ is an empty polyhedron. Hence assuming

$$A\bar{x} \leq b \qquad \text{for some } \bar{x} \in \mathbf{K}^n \tag{4.1}$$

must lead to a contradiction.

By monotonicity of addition and multiplication with nonnegative scalars, it is immediate that (4.1) implies

$$\forall\, u \in \mathbf{K}_+^m \qquad u^T A\bar{x} \leq u^T b. \tag{4.2}$$

Since we do not know anything about $\bar{x}$, we will in general not be able to decide whether $u^T A\bar{x} \leq u^T b$ holds for a given $u \in \mathbf{K}_+^m$ or not. Therefore we are interested in those inequalities which are independent of $\bar{x}$, i.e. we consider those $u \in \mathbf{K}_+^m$, which satisfy $u^T A = 0$. This leads to

$$\forall\, u \in \mathbf{K}_+^m \quad (u^T A = 0 \Rightarrow u^T b \geq 0). \tag{4.3}$$

Thus if we find $u \in \mathbb{K}_+^m$ such that $u^T A = 0$ but $u^T b < 0$, we have a contradiction to (4.1). FARKAS' Lemma now also proves the converse, i.e. if we can **not** find $u \in \mathbb{K}_+^m$ such that $u^T A = 0$ and $u^T b < 0$, then statement (4.1) must be true.

Theorem 4.1 (FARKAS' Lemma) *Given $A \in \mathbb{K}^{m\times n}$ and $b \in \mathbb{K}^m$, then either*

a) $\exists x \in \mathbb{K}^n,\ Ax \leq b$

or

b) $\exists u \in \mathbb{K}_+^m,\ u^T A = 0,\ u^T b < 0$

but not both.

Instead of proving Theorem 4.1, we will prove an equivalent statement in Section 4.3 (see Proposition 4.5).

Using scalar multiplication, condition b) of Theorem 4.1 is equivalent to

b') $\exists u \in \mathbb{K}_+^m,\ A^T u = 0,\ b^T u = -1.$

Thus we may restate Theorem 4.1 as follows:

Corollary 4.2 *Given $A \in \mathbb{K}^{m\times n}$ and $b \in \mathbb{K}^m$, then either*

a) $P = P(A, b) \neq \emptyset$

or

b) $P' = P^{=}\left(\begin{pmatrix} A^T \\ b^T \end{pmatrix}, \begin{pmatrix} 0 \\ -1 \end{pmatrix}\right) \neq \emptyset$

but not both.

□

4.2 Homogenization

In Chapter 1 we defined polyhedral cones to be special polyhedra with right hand side $b = 0$. In this section we shall show that they are in a sense as general as polyhedra. Indeed we shall construct for a given polyhedron $P \subseteq \mathbb{K}^n$ a polyhedral cone C which still contains all information about P. Needless to say that polyhedral cones usually are much easier to handle than general polyhedra. Therefore this construction principle, the socalled "homogenization", together with the corresponding inverse operation, the socalled "dehomogenization" are useful tools in polyhedral theory.

Before we come to a formal definition of homogenization, let us discuss what we have in mind. Since a polyhedron is the intersection of finitely many affine halfspaces, while a polyhedral cone is the intersection of finitely many (linear) halfspaces, we have to look for a construction which transforms affine halfspaces into linear halfspaces. Such a construction is wellknown in linear algebra, where it is used to pass from affine spaces to projective ones: If $A \subseteq \mathbf{K}^n$ is an affine subspace, then $L := \left\{\lambda\binom{x}{1} \mid x \in A, \lambda \in \mathbf{K}\right\}$ is a linear subspace of $\mathbf{K}^{n+1}$.

Following this idea, we might define the "homogenization" of a given polyhedron $P \subset \mathbf{K}^n$ to be the conic hull of P in $\mathbf{K}^{n+1}$ translated at level 1 (cf. Figure 4.1).

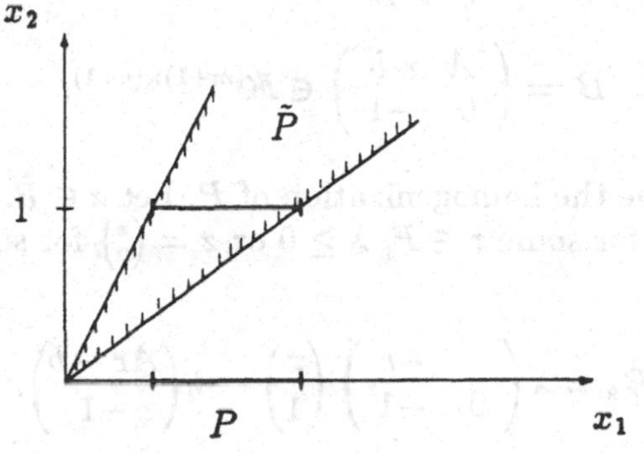

Figure 4.1

Hence the homogenization of P would be the set $\tilde{P} = \left\{\lambda\binom{x}{1} \mid x \in P, \lambda \in \mathbf{K}_+\right\}$. Unfortunately, this set is in general not a polyhedral cone, it even may not be closed. For example, if $P = \{x \mid x \geq 0\} \subseteq \mathbf{R}$, then $\tilde{P} = \left\{\binom{x}{y} \mid x \geq 0, y > 0\right\} \cup \binom{0}{0}$, which is not a polyhedral cone. This example indicates that we have to add the set $\left\{\binom{x}{0} \mid x \in rec\, P\right\}$.

Definition 4.3 *Let $P \subseteq I\!\!K^n$ be a polyhedron. Then the set*

$$P := \left\{\lambda\binom{x}{1} \,\middle|\, x \in P, \lambda \in I\!\!K_+\right\} \cup \left\{\binom{x}{0} \,\middle|\, x \in rec\, P\right\}$$

is called the **homogenization** *of P. Usually, we will shorten the notation by writing*

$$\tilde{P} = \mathbb{K}_+ \binom{P}{1} + \binom{rec\, P}{0}.$$

Instead of translating P to level 1 we might have chosen any other level $\tau \in \mathbb{K} \setminus \{0\}$. Sometimes it is convenient to choose $\tau = -1$, and therefore we also introduce the **negative homogenization** *of P, defined by*

$$\tilde{P} = \mathbb{K}_+ \binom{P}{-1} + \binom{rec\, P}{0}.$$

The following result provides an explicit description of $\tilde{P}$ in terms of a system of linear inequalities (thereby showing that $\tilde{P}$ in fact is a polyhedral cone):

Proposition 4.4 *Let $P = P(A, b)$ be a polyhedron in $\mathbb{K}^n$. Then the homogenization of P is $P(B, 0)$, where*

$$B := \begin{pmatrix} A & -b \\ 0 & -1 \end{pmatrix} \in \mathbb{K}^{(m+1)\times(n+1)}.$$

Proof. Let $\tilde{P}$ be the homogenization of P. Let $z \in \tilde{P}$. By definition, either $z = \lambda\binom{x}{1}$ for some $x \in P$, $\lambda \geq 0$ or $z = \binom{x}{0}$ for some $x \in rec\, P$. In the first case,

$$Bz = \lambda \begin{pmatrix} A & -b \\ 0 & -1 \end{pmatrix} \binom{x}{1} = \lambda \binom{Ax - b}{-1}.$$

Since $Ax \leq b$ and $\lambda \geq 0$, it follows that $Bz \leq 0$, i.e. $z \in P(B, 0)$. The second case is immediate from Lemma 1.7, thus $\tilde{P} \subseteq P(B, 0)$.

Now let $z = \binom{x}{z_{n+1}} \in P(B, 0)$. Then $0x + (-1)z_{n+1} \leq 0$, hence $z_{n+1} \geq 0$. If $z_{n+1} > 0$, $z = \lambda\binom{x}{1}$ for some $\lambda > 0$. Since $Bz \leq 0$, we must have $B\binom{x}{1} \leq 0$, that is $Ax \leq b$, showing that $z \in \tilde{P}$. If $z_{n+1} = 0$, then $Ax \leq 0$, i.e. $x \in rec\, P$. Thus we have shown that $P(B, 0) \subseteq \tilde{P}$. □

It is left to the reader to prove a similar result for the negative homogenization of P.

As an application of this construction , we will now "homogenize" the FARKAS Lemma (Theorem 4.1):

Theorem 4.5 *Let $A \in \mathbb{K}^{m\times n}$, $b \in \mathbb{K}^m$ and $B := \begin{pmatrix} A & -b \\ 0 & -1 \end{pmatrix}$. Then either*

a) $\exists\, z \in \mathbb{K}^{n+1}\,,\; Bz \leq 0\,,\; B_{(m+1)\cdot}z < 0$

or

b) $\exists\, u \in \mathbb{K}_{+}^{m+1}\,,\; u^T B = 0\,,\; u_{m+1} > 0$

but not both.

Proof. We shall only show that Theorem 4.5 and Theorem 4.1 imply each other. The proof of both statements will be given in Section 4.3. Clearly Theorem 4.5 a) is equivalent to "$\exists\, z \in \mathbf{K}^{n+1}\,,\; Bz \leq 0$, $B_{(m+1)\cdot}z = -1$" or "$\exists\, z \in \mathbf{K}^{n+1}\,,\; Bz \leq 0\,,\; z_{n+1} = 1$." Again this can be reformulated as "$\exists\, \binom{x}{1} \in \mathbf{K}^{n+1}\,,\; B\binom{x}{1} \leq 0$" or as "$\exists\, x \in \mathbf{K}^n : Ax \leq b$" which is 4.1 a). In the same way, we see that 4.5 b) is equivalent to 4.1 b).

□

Theorem 4.5 is just a restatement of Theorem 4.1, and there seems to be nothing interesting about it. However, we are now prepared to change our point of view. Note that the set $\{Bz \mid z \in \mathbf{K}^{n+1}\} = im\, B$ is a vectorspace in $\mathbf{K}^{m+1}$ and that $\{u \mid u^T B = 0\} = ker\, B^T$ is its orthogonal complement. Thus the FARKAS Lemma can be seen as a theorem relating pairs of orthogonal vectorspaces. We will emphasize this point of view in the next section.

4.3 Linearization

As we used homogenization to pass from polyhedra to polyhedral cones, we will use "linearization", to pass from polyhedral cones to linear vectorspaces. However, "linearization" is not really a "technique" in the usual sense, but rather a "change of viewpoint". Suppose we are given a polyhedral cone $C = P(B, 0)$, $B \in \mathbf{K}^{(m+1)\times(n+1)}$. Instead of looking at $C \subseteq \mathbf{K}^{n+1}$, we may as well apply the transformation $B : \mathbf{K}^{n+1} \to \mathbf{K}^{m+1}$. Anything happening to C in $\mathbf{K}^{n+1}$ will then correspond to something happening in $im\, B \subseteq \mathbf{K}^{m+1}$. Any "polyhedral statement" concerning C will correspond to a "linear subspace statement" concerning $L = im\, B$.

As an example, how we can use "linearization" to translate statements about polyhedral cones into equivalent statements about vectorspaces, we shall restate the FARKAS Lemma as follows:

Theorem 4.6 *Let $L \leq \mathbb{K}^{m+1}$ be a subspace of $\mathbb{K}^{m+1}$. Then either*

a) $\exists y \in L, y \geq 0, y_{m+1} > 0$

or

b) $\exists u \in L^{\perp}, u \geq 0, u_{m+1} > 0$

but not both.

Proof. We shall first show that Theorem 4.6 is equivalent to Theorem 4.5 and will then give a proof of Theorem 4.6. Given $A \in \mathbb{K}^{m \times n}$, $b \in \mathbb{K}^m$, and $B = \begin{pmatrix} A & -b \\ 0 & -1 \end{pmatrix}$ as in Theorem 4.5, we set $L := im\, B$, thus $L^{\perp} = ker\, B^T$. Now Theorem 4.6 states, that either

a) $\exists y = Bz, y \geq 0, \quad y_{m+1} > 0$

or

b) $\exists u \geq 0, \ u^T B = 0, u_{m+1} > 0$

but not both. This is exactly the content of Theorem 4.5.

To show that Theorem 4.6 follows from Theorem 4.5, let L be an arbitrary subspace of $\mathbb{K}^{m+1}$. Since L and $L^{\perp}$ together span $\mathbb{K}^{m+1}$, it cannot happen that both L and $L^{\perp}$ are contained in $\mathbb{K}^m \times \{0\}$. Due to the symmetry of the conditions a) and b) in Theorem 4.6, we may assume w.l.o.g. that L is not contained in $\mathbb{K}^m \times \{0\}$. Hence L can be written as $L = im\, B$, where $B \in \mathbb{K}^{(m+1)\times(n+1)}$ has the form $\begin{pmatrix} A & -b \\ 0 & -1 \end{pmatrix}$. The claim now follows immediately from Theorem 4.5.

□

Proof. (Theorem 4.6) Obviously, the two alternatives a) and b) cannot hold simultaneously. For, if y is as in a) and u is as in b), then $y^T u > 0$, contradicting the definition of $L^{\perp}$. Therefore, it is sufficient to show that for every subspace $L \leq \mathbb{K}^{m+1}$, there exists $y \in L$ and $u \in L^{\perp}$ such that $y \geq 0$, $u \geq 0$ and $y_{m+1} + u_{m+1} > 0$. This is clearly true if $m = 0$, thus we may proceed by induction on m. Assume that $m \geq 1$, and let A be a matrix such that $L = ker\, A$. Now delete the first column of A and denote the resulting matrix by $\hat{A}$.

Then by our induction hypothesis, there exist $\hat{y} = (\hat{y}_2, \ldots, \hat{y}_{m+1}) \in ker\, \hat{A}$ and $\hat{u} = (\hat{u}_2, \ldots, \hat{u}_{m+1}) = \hat{w}^T \hat{A} \in im\, \hat{A}^T$, such that $\hat{y} \geq 0$, $\hat{u} \geq 0$ and $\hat{y}_{m+1} + \hat{u}_{m+1} > 0$. Considering $\bar{y} := (0, \hat{y}) \in L$ and $\bar{u} := \hat{w}^T A \in L^{\perp}$, we see that

(1) There exists $\bar{y} \in L$, $\bar{u} \in L^{\perp}$ such that $\bar{y} \geq 0$, $(\bar{u}_2, \ldots, \bar{u}_{m+1}) \geq 0$ and $\bar{y}_{m+1} + \bar{u}_{m+1} > 0$.

By symmetry, our induction hypothesis implies:

(2) There exist $\bar{\bar{y}} \in L$, $\bar{\bar{u}} \in L^{\perp}$ such that $(\bar{\bar{y}}_2 \ldots, \bar{\bar{y}}_{m+1}) \geq 0$, $\bar{\bar{u}} \geq 0$, and $\bar{\bar{y}}_{m+1} + \bar{\bar{u}}_{m+1} > 0$.

If $\bar{y}$ and $\bar{u}$ are as in (1) and $\bar{u}_1 \geq 0$, we are done. Similarily, if $\bar{\bar{y}}$ and $\bar{\bar{u}}$ are as in (2) and $\bar{\bar{y}}_1 \geq 0$, we are done. However at least one of these must occur, for if $\bar{u} \in L^{\perp}$ and $\bar{\bar{y}} \in L$ are as in (1) and (2) resp. then $\bar{u} \cdot \bar{\bar{y}} = 0$, implying $\bar{u}_1 \cdot \bar{\bar{y}}_1 = -\bar{u}_2 \cdot \bar{\bar{y}}_2 - \ldots - \bar{u}_{m+1} \cdot \bar{\bar{y}}_{m+1} \leq 0$.

□

Clearly there is nothing special about the $(m+1)$-th coordinate in $\mathbb{K}^{m+1}$ and we can substitute $m+1$ by any $i \in \{1, \ldots, m+1\}$ in Theorem 4.6. This proves

Corollary 4.7 *If $L \leq \mathbb{K}^m$ is a subspace and $i \in \{1, \ldots, m\}$, then either*

a) $\exists\, y \in L$, $y \geq 0$, $y_i > 0$

or

b) $\exists\, u \in L^{\perp}$, $u \geq 0$, $u_i > 0$

but not both.

□

Example 4.8 *Consider the 3-dimensional vectorspace $\mathbb{K}^3$ and let L be a (hyper-) plane through the origin in $\mathbb{K}^3$. Then Corollary 4.7 states, that for every $i \in \{1, 2, 3\}$ either L or one of its normal vectors meets the intersection of the open halfspace $\{x \in \mathbb{K}^3 \mid x_i > 0\}$ with the first octant $\{x \in \mathbb{K}^3 \mid x \geq 0\}$.*

4.4 Delinearization

Delinearization is the reverse of linearization. Thus delinearization may be used to derive results about polyhedral cones from results about linear subspaces. As we have seen, the "polyhedral cone" version of FARKAS' Lemma may be derived from the "linear subspace" version by means of delinearization. More precisely, we obtained Theorem 4.5 from Theorem 4.6 by writing L and $L^{\perp}$ as $L = im\, B$ and $L^{\perp} = ker\, B^T$. In this section we will present another "polyhedral cone" version of FARKAS' Lemma which can be obtained from Theorem 4.6 in the very same way, this time, however, interchanging the roles of B and B^T, i.e. by writing $L = ker\, B$ and $L^{\perp} = im\, B^T$. Here is what one gets:

Theorem 4.9 *Let A be a finite subset of $\mathbb{K}^n$ and $b \in \mathbb{K}^n$. Then either*

a) $b \in cone\, A$

or

b) $\exists z \in \mathbb{K}^n,\ z^T A \leq 0,\ z^T b > 0$

but not both.

Proof. Let $A \in \mathbf{K}^{m \times n}$ and set $B := (A, -b)$ and $L := ker\, B$. Hence $L^\perp = im\, B^T$. From Theorem 4.6 we get that either

a') $\exists y \in \mathbf{K}^{m+1},\ (A, -b)y = 0,\ y \geq 0,\ y_{m+1} > 0$

or

b') $\exists z \in \mathbf{K}^n,\ z^T(A, -b) \geq 0,\ z^T(-b) > 0$

but not both.

Clearly, in a') we may require $y_{m+1} = 1$ instead of $y_{m+1} > 0$. Thus, a') is equivalent to

a") $\exists x \in \mathbf{K}^n,\ Ax = b,\ x \geq 0,$

which is the same as Theorem 4.9 a).

Replacing z by $-z$, we see that b') is equivalent to Theorem 4.9 b). Thus we have shown that Theorem 4.6 implies Theorem 4.9.

To prove the converse, let $L \leq \mathbf{K}^{m+1}$ be an arbitrary subspace of $\mathbf{K}^{m+1}$. Clearly, $L = ker\, B$ for some matrix $B \in \mathbf{K}^{n \times (m+1)}$ and every such matrix can be written in the form $(A, -b)$ with $A \in \mathbf{K}^{n \times m}$, $b \in \mathbf{K}^n$. Now it is imediate that Theorem 4.9 implies Theorem 4.6.

□

Given a subset M of $\mathbf{K}^n$ and a vector $b \in \mathbf{K}^n$, we say that b can be **separated from M** if there exists a closed linear halfspace $\overline{H^+}$ of $\mathbf{K}^n$ such that $M \subseteq \overline{H^+}$ but $b \notin \overline{H^+}$. Using this terminology, Theorem 4.9 can be stated as follows:

either

a) $b \in cone\, A$

or

b) b can be separated from A

but not both.

By definition, b cannot be separated from A if and only if it is contained in every closed halfspace containing A, i.e. if and only if it is contained in the intersection of all the closed halfspaces containing A. Thus we get as a corollary from 4.9:

Corollary 4.10 *Let $A \subseteq \mathbb{K}^n$ be a finite set. Then cone A is the intersection of the closed halfspaces containing A.*

□

Actually, a more general result (which we will not prove) states that every closed cone in $\mathbb{K}^n$ is representable as an intersection of halfspaces. We shall see later, that if A is finite, then *cone* A is even representable as the intersection of a finite number of halfspaces — hence, that *cone* A is a polyhedral cone.

4.5 Dehomogenization

To end with our presentation of the Farkas Lemma, we "dehomogenize" Theorem 4.9. Thus, coming back to general polyhedra, which have been our starting point, we will prove that the convex hull of a finite set $A \subseteq \mathbb{K}^n$ is representable as the intersection of affine halfspaces.

Theorem 4.11 *Let $V, E \subseteq \mathbb{K}^n$ be finite sets and $b \in \mathbb{K}^n$. Then either*

a) $b \in conv\, V + cone\, E$

or

b) $\exists \binom{c}{c_0} \in \mathbb{K}^{n+1}$ *s.t.* $cv \leq c_0 \; \forall\, v \in V,\; ce \leq 0 \; \forall\, e \in E,\; cb > c_0$

but not both.

Proof. In order to apply Theorem 4.9, we first "homogenize" the set $P := conv\, V + cone\, E$, i.e. we set

$$A := \begin{pmatrix} V & E \\ -\mathbf{1}^T & 0 \end{pmatrix}.$$

Hence $b \in P$ if and only if $\binom{b}{-1} \in cone\, A$. Now Theorem 4.9 states, that either

a') $\binom{b}{-1} \in cone\, A$

or

b') $\exists z \in \mathbb{K}^{n+1}, z^T A \leq 0, z^T \binom{b}{-1} > 0.$

Since a) is equivalent to a'), it suffices to show that b) is equivalent to b'). However, writing $z = \binom{c}{c_0} \in \mathbb{K}^{n+1}$, we see immediately that b) and b') are the same.

□

We have proved, that Theorem 4.11 follows from Theorem 4.9. However, setting $V = \{0\}$, we see that Theorem 4.9 is only a special case of Theorem 4.11. Thus again, Theorem 4.11 is a restatement of the FARKAS Lemma.

As in the case of polyhedral cones, Theorem 4.11 shows that a vector $b \in \mathbb{K}^n$ is in $P := conv\, V + cone\, E$ if and only if it is contained in every affine halfspace $\{x \mid cx \leq c_0\}$ that contains P. We therefore get the following

Corollary 4.12 *Let $V, E \subseteq \mathbb{K}^n$. Then $conv\, V + cone\, E$ is representable as an intersection of affine halfspaces.*

□

Here again, as we will see, a finite number of halfspaces suffices to represent $conv\, V + cone\, E$, i.e. that $cone\, V + cone\, E$ is a polyhedron.

There are various extensions of the FARKAS Lemma in different directions. For instance Theorem 4.1 can be stated in a more general frame using subsets $S, U \subseteq \mathbb{K}$, i.e.

4.13

Either

a) $\exists\, x \in \mathbb{K}^n \quad b - Ax \in S^m$

or

b) $\exists\, u \in U^m \quad u^T A = 0,\ ub \notin S$

Clearly $S = \mathbb{K}_+$ and $U = \mathbb{K}_+$ is the original FARKAS Lemma and $S = \{0\}$, $U = K$ is a well known theorem about the consistency of linear equations. But (4.13) is also true for $S = U = \mathbb{Z}$ (cf. A. Bachem and R.v. Randow [8]. More difficult is the case $S = U = \mathbb{Z}_+$ for which (4.13) becomes false. See D. Crystal [54] for a treatment of this case.

4.6 Further Reading

FARKAS' Lemma dates back to the beginning of this century, cf.

J. FARKAS *Theorie der einfachen Ungleichungen*, Journal für Reine und Angewandte Mathematik **124** (1902), pp. 1–27, although it had been prepared already by Fourier, cf.

J.-B. FOURIER *Solution d'une Question Particulière du Calcul des Inégalités*, Oevres II (1836), pp. 317–328.

The proof technique based on this approach, the socalled FOURIER – MOTZKIN Elimination is applied, e.g. in

J. STOER, CH. WITZGALL *Convexity and Optimization in Finite Dimension I*, Springer (1970),

to derive FARKAS' Lemma. There one can find also the more general theorem about representing closed convex sets by intersections of affine halfspaces.

Chapter 5
Oriented Matroids

In Chapter 4 we learned about at least five versions of FARKAS' Lemma. Is there any one which is more "natural" than the others? We propagate that the "linear subspace version" (Theorem 4.6) is the most natural one: It is both simple and "selfdual", i.e. both alternatives a) and b) arise from each other by simply replacing L by $L^\perp$. If we agree that Theorem 4.6 is the most natural way to state FARKAS' Lemma, the next question that arises is: Are dual pairs of vector spaces the most natural structures for stating Theorem 4.6? More precisely, given two sets of vectors, say, $\mathcal{S}$ and $\mathcal{S}'$ such that an analogue of Theorem 4.6 holds (with L and $L^\perp$ replaced by $\mathcal{S}$ an $\mathcal{S}'$, resp.), is it true then that $\mathcal{S}$ and $\mathcal{S}'$ give rise to a dual pair of vector spaces in some way? As one might guess, the answer is no (for otherwise we would probably not have written this book). In fact, a systematic analysis of those properties of vector spaces which make L and $L^\perp$ satisfy FARKAS' Lemma will lead us to discover more general structures, called "oriented matroids". These are, as we will see, the most general (and hence the most simple or "natural") structures satisfying an analogue of FARKAS' Lemma.

After all, as it happens so often, the whole theory becomes much more interesting than the original question from which it arose. Thus we will have no time to show that there actually do exist oriented matroids that do **not** correspond to vector spaces until Chapter 9.

5.1 Sign Vectors

A sign vector is a vector whose components are signs, i.e. either "+" or "−" or "0". If $x \in \mathbb{K}^n$, then $\sigma(x)$ shall denote the sign vector that corresponds to x in the obvious way. Throughout, sign vectors will be denoted by upper case letters. Thus $X = \sigma(x)$ means $X_i = +(-, 0)$

if and only if $x_i > 0 (< 0, = 0)$. Using this notation, we may restate Corollary 4.7 as follows:

Let $\mathcal{S}$ and $\mathcal{S}'$ denote the sets of sign vectors corresponding to vectors in L and L', resp. Then for every i either

a) $\exists X \in \mathcal{S}, X_i = +, X \geq 0$

or

b) $\exists Y \in \mathcal{S}', Y_i = +, Y \geq 0$

but not both.

Here, of course, $X \geq 0$ shall mean that X has a "+"- or a "0"-entry in every coordinate. For notational convenience, sign vectors will usually be indexed by elements of a finite set E, and they will always be written as row vectors.

Definition 5.1 *Let $E \neq \emptyset$ be finite. An element $X \in \{+,-,0\}^E$ is called a* **sign vector** *(on E). The set of all sign vectors on E is denoted by*

$$2^{\pm E} := \{+,-,0\}^E.$$

The symbols "0", "+" and "−" are referred to as **signs**. *We say that "+" is the* **opposite** *sign of "−" and vice versa. For a sign vector $X \in 2^{\pm E}$ we introduce its*

	positive part	X^+	:=	$\{e \in E \mid X_e = +\}$,
	negative part	X^-	:=	$\{e \in E \mid X_e = -\}$,
	zero part	X^0	:=	$\{e \in E \mid X_e = 0\}$,
and its	**support**	$supp X$	:=	$X^+ \cup X^-$

We denote by $-X$ the sign vector which is **opposite** *to X, i.e. whose negative (positive) part equals the positive (negative) part of X. Clearly this notion also applies to sets of sign vectors, thus for $\mathcal{S} \subseteq 2^{\pm E}$, $-\mathcal{S}$ denotes the set of opposite sign vectors. If X has no negative components on $A \subseteq E$ (i.e. $X^- \cap A = \emptyset$), we say that X is* **nonnegative** *on A, denoted by $X_A \geq 0$. $X_A \leq 0$ and $X_A = 0$ are defined similarily. We will usually write $X \geq 0$ and $X \leq 0$ instead of $X_E \geq 0$ and $X_E \leq 0$, resp.*

If $\mathcal{S}$ and $\mathcal{S}'$ are two sets of sign vectors, then the pair $(\mathcal{S}, \mathcal{S}')$ is said to have the FARKAS **Property**, *if*

(FP) $\forall e \in E$ *either*

a) $\exists X \in \mathcal{S}$, $e \in supp X$ *and* $X \geq 0$

or

b) $\exists Y \in \mathcal{S}'$, $e \in supp\,Y$ *and* $Y \geq 0$,

but not both.

Example 5.2 *Let* $\sigma : I\!\!K^E \to 2^{\pm E}$ *denote the "forgetting magnitudes-map", which associates to every* $x \in I\!\!K^E$ *the corresponding sign vector* $X = \sigma(x)$, *and let* $\mathcal{S} = \sigma(L)$ *and* $\mathcal{S}' = \sigma(L^\perp)$ *for some pair of orthogonal subspaces. Then* $(\mathcal{S}, \mathcal{S}')$ *has the* FARKAS *Property (cf. Corollary 4.7).*

For example, let

$$A := \begin{pmatrix} 0 & 1 \\ 1 & 0 \\ 1 & -1 \\ 1 & -2 \end{pmatrix}$$

and let $L := im\,A$. *Let* $x \in I\!\!K^2$ *and* $y = Ax \in L$. *Then, in a sense, the sign vector* $Y = \sigma(y)$ *reflects the position of* x *with respect to the hyperplanes (lines) whose normal vectors are the rows* $A_{i.}$. *In fact,* $Y_i > 0$ (< 0), *according to whether* $A_{i.}$ *and* x *are on the same (opposite) side of the line* $l_i := \{z \in I\!\!K^2 \mid A_{i.}z = 0\}$ *(cf. Figure 5.1).*

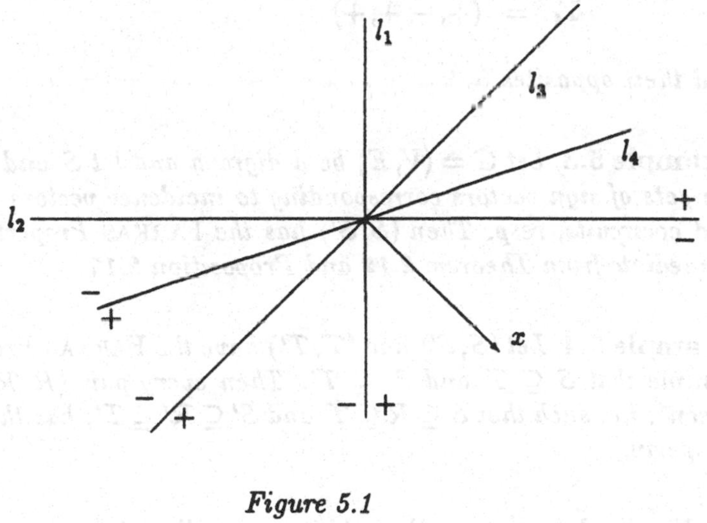

Figure 5.1

If $x \in I\!\!K^2$ *is as indicated in Figure 5.1, then the sign vector* Y *corresponding to* $y = Ax$ *is* $Y = (+, -, +, +)$. *Moving the vector* x *in the plane* $I\!\!K^2$, *we get a complete list of all sign vectors in* $\mathcal{S} := \sigma(L)$.

There are precisely 17 of them:

$$\begin{array}{rclrcl} X_0 & = & (0,0,0,0) & X_5 & = & (+,+,0,-) \\ X_1 & = & (+,0,+,+) & X_6 & = & (+,+,-,-) \\ X_2 & = & (+,+,+,+) & X_7 & = & (0,+,-,-) \\ X_3 & = & (+,+,+,0) & X_8 & = & (-,+,-,-) \\ X_4 & = & (+,+,+,-) & & & \end{array}$$

and their opposites.

Similarily, one may determine $\mathcal{S}' = \sigma(L^{\perp})$. *Note that, e.g.* $(1,-1,1,0)^T$ *and* $(1,0,-1,1)^T$ *form a basis of* $L^{\perp}$, *hence*

$$L^{\perp} = im \begin{pmatrix} 1 & 1 \\ -1 & 0 \\ 1 & -1 \\ 0 & 1 \end{pmatrix}.$$

The sign vectors in $\mathcal{S}'$ *are*

$$\begin{array}{rclrcl} Y_0 & = & (0,0,0,0) & Y_5 & = & (+,-,0,+) \\ Y_1 & = & (0,-,+,-) & Y_6 & = & (+,-,-,+) \\ Y_2 & = & (+,-,+,-) & Y_7 & = & (+,0,-,+) \\ Y_3 & = & (+,-,+,0) & Y_8 & = & (+,+,-,+) \\ Y_4 & = & (+,-,+,+) & & & \end{array}$$

and their opposites.

Example 5.3 *Let* $G = (V, E)$ *be a digraph and let* $\mathcal{S}$ *and* $\mathcal{S}'$ *denote the sets of sign vectors corresponding to incidence vectors of circuits and cocircuits, resp. Then* $(\mathcal{S}, \mathcal{S}')$ *has the* FARKAS *Property. This is immediate from Theorem 2.12 and Proposition 2.17.*

Example 5.4 *Let* $(\mathcal{S}, \mathcal{S}')$ *and* $(\mathcal{T}, \mathcal{T}')$ *have the* FARKAS *Property and assume that* $\mathcal{S} \subseteq \mathcal{T}$ *and* $\mathcal{S}' \subseteq \mathcal{T}'$. *Then every pair* $(\mathcal{R}, \mathcal{R}')$ *"in between", i.e. such that* $\mathcal{S} \subseteq \mathcal{R} \subseteq \mathcal{T}$ *and* $\mathcal{S}' \subseteq \mathcal{R}' \subseteq \mathcal{T}'$, *has the* FARKAS *Property.*

Finally, let us note, that there are really trivial examples, too. Let $\mathcal{S}$ consist of a single sign vector $Y = (+, \cdots, +)$, and let $\mathcal{S}' = \emptyset$. Then, obviously, $(\mathcal{S}, \mathcal{S}')$ has the FARKAS Property. The existence of such degenerate examples indicates that we should impose further conditions, in order to exhibit interesting structures.

5.2 Minors

As indicated in the last paragraph, simply requiring the FARKAS Property to hold for a pair $(\mathcal{S}, \mathcal{S}')$ does not lead to an interesting theory. So what other "structural" properties of vector spaces can be translated to sign vectors? One of the most important characteristics of vector spaces is that the intersection of two (or more) vector spaces is a vector space again. In particular, if $L \leq \mathbf{K}^E$ is a subspace, and $e \in E$, then

$$\{x \in L \mid x_e = 0\}$$

is a subspace again. If $L = ker\,A$, for some matrix A whose columns are indexed by E, this operation corresponds to deleting column e from A. Thus let $A \setminus e$ denote the matrix obtained from A by deleting column e. Then

$$L \setminus e := ker(A \setminus e) \leq \mathbf{K}^{E \setminus e}$$

is a subspace of $\mathbf{K}^{E\setminus e}$. In terms of L, this can be described as follows:

$$L \setminus e = \{\tilde{x} \in \mathbf{K}^{E\setminus e} \mid \exists x \in L, x_e = 0 \quad \text{and} \quad x = \tilde{x} \quad \text{on} \quad E \setminus e\}.$$

The dual of $L \setminus e$ is $(L \setminus e)^{\perp} = im\,(A \setminus e)^T$. This may be explicitely described as

$$(L \setminus e)^{\perp} = \{\tilde{y} \in \mathbf{K}^{E\setminus e} \mid \exists y \in L^{\perp},\ y_e = * \ \text{ and } \ y = \tilde{y} \text{ on } E \setminus e\}.$$

Here, obviously, "$y_e = *$" means "y_e is arbitrary". Thus, in some sense, the dual of $L \setminus e$ equals "$L^{\perp}$ modulo coordinate e". For this reason, the $(L \setminus e)^{\perp}$ will also be denoted by $L^{\perp}/e$. We will say that $L \setminus e$ is obtained from L by **deleting** e and $L^{\perp}/e$ is obtained from $L^{\perp}$ by **contracting** e .

Example 5.5 *Recall the corresponding notions for graphs. Let $G = (V, E)$ be a graph with circuitspace $L \leq \mathbb{K}^E$ and cocircuitspace $L^{\perp} \leq \mathbb{K}^E$. Then $G \setminus e$ and G/e denote the graphs obtained from G by deleting resp. contracting the edge $e \in E$. Obviously, the circuits in $G \setminus e$ are precisely those circuits of G which do not pass through e. From this it is immediate that $L \setminus e$ is the circuitspace of $G \setminus e$, and, consequently, its dual $L^{\perp}/e$ is the cocircuitspace of $G \setminus e$. Similarily, the circuitspace of G/e is easily seen to be L/e and hence the cocircuitspace of G/e is $L^{\perp} \setminus e$.*

Perhaps the duality between the two operations deletion and contraction is best understood in the case of planar graphs. Assume G is planar and $G^\perp$ is its dual. Thus L and $L^\perp$ are the circuitspaces of G and $G^\perp$, resp. Assume that $e \in E$ is contained in the boundary of two regions R_1 and R_2 in some embedding of G. Then deleting e amounts to joining the two regions R_1 and R_2 to a single region R. In the dual graph $G^\perp$, this amounts to identifying the two vertices corresponding to R_1 and R_2 and removing the edge e^*. Thus in fact, $G^\perp/e^*$ is the dual of $G\setminus e$. As we have seen above, $L\setminus e$ is the circuitspace of $G\setminus e$ and hence $L^\perp/e$ is the circuitspace of $G^\perp/e^*$.

Of course, instead of deleting single elements of E, we may also delete a subset $I \subseteq E$ by deleting each $e \in I$. Furthermore, we may both delete some elements in L, contract some others and perform the corresponding dual operations on $L^\perp$ in order to construct new pairs of dual subspaces: Let I and J denote disjoint subsets of E. Then

$$\begin{aligned} L\setminus I/J &= \{\tilde{x} \in \mathbb{K}^{E\setminus(I\cup J)} \mid \exists\, x \in L,\ x_I = 0,\ x_J = * \\ &\qquad \text{and } x = \tilde{x} \text{ on } E\setminus(I\cup J)\} \end{aligned}$$

is called a **minor** of L. Of course, "$x_I = 0$" reads as "$x_i = 0\ \forall i \in I$" and "$x_J = *$" means "x_j is arbitrary $\forall j \in J$".

The dual of $L\setminus I/J$ is obviously given by $L^\perp/I\setminus J$. The dual pair of subspaces $(L\setminus I/J, L^\perp/I\setminus J)$ will be called a **minor** of $(L, L^\perp)$. This notion carries over to sign vectors.

Definition 5.6 *Let $\mathcal{S} \subseteq 2^{\pm E}$, and let I and J denote disjoint subsets of E. Then*

$$\begin{aligned} \mathcal{S}\setminus I/J &= \{\tilde{X} \in 2^{\pm E\setminus(I\cup J)} \mid \exists\, X \in \mathcal{S},\ X_I = 0,\ X_J = *, \\ &\qquad X = \tilde{X} \ \text{ on } E\setminus(I\cup J)\} \end{aligned}$$

is called a **minor** *of $\mathcal{S}$ (obtained by* **deleting** *I and* **contracting** *J). If $\mathcal{S}, \mathcal{S}'$ are sets of sign vectors on E, then $(\mathcal{S}\setminus I/J, \mathcal{S}'\setminus J/I)$ is called a* **minor** *of $(\mathcal{S}, \mathcal{S}')$.*

Note that the elements of E may be deleted or contracted in any order, so we are allowed to ommit brackets or write $S\,/\,J\setminus I$ instead of $\mathcal{S}\setminus I\,/\,J$. It is perhaps a good idea to visualize the minor operation as indicated below

$$\begin{aligned} \tilde{X} &= (+,\dots,+,0,\dots,0,-,\dots,-) \in \mathcal{S}\setminus I/J \\ \text{iff } \exists X &= (\underbrace{+,\dots,+,0,\dots,0,-,\dots,-}_{E\setminus(I\cup J)},\underbrace{0,\dots,0}_{I},\underbrace{*,\dots,*}_{J}) \in \mathcal{S} \end{aligned}$$

There is one more operation by which one may derive "new" subspaces from a given one. This is a really trivial one, called **reorientation**: Given $L \leq \mathbb{K}^E$, and $I \subseteq E$, then

$$_IL := \{\tilde{x} \in \mathbb{K}^E \mid \exists\, x \in L,\ x_I = -\tilde{x}_I,\ x_{E\setminus I} = \tilde{x}_{E\setminus I}\}$$

is again a subspace of $\mathbb{K}^E$.

For example, if L is the circuit space of a digraph $G = (V, E)$, then ${}_IL$ is the circuit space of the graph ${}_IG$, which arises from G by **reorienting** the edges $e \in I$, i.e. replacing every such edge $e = (u, v)$ by $-e = (v, u)$.

Definition 5.7 *Let $\mathcal{S}, \mathcal{S}' \subseteq 2^{\pm E}$ and $I \subseteq E$. Then*

$$_I\mathcal{S} := \{\tilde{X} \in 2^{\pm E} \mid \exists X \in \mathcal{S},\ X_I = -\tilde{X}_I,\ X_{E \setminus I} = \tilde{X}_{E \setminus I}\}$$

is called the **reorientation** *of $\mathcal{S}$ on I. $({}_I\mathcal{S}, {}_I\mathcal{S}')$ is called the* **reorientation** *of $(\mathcal{S}, \mathcal{S}')$ on I.*

Note that the minor–operation commutes with reorientation, so we are again allowed to ommit brackets in expressions like ${}_I\mathcal{S} \setminus K/J$.

5.3 Oriented Matroids

In the last section we learned of some operations by means of which we may derive new dual pairs of vector spaces from a given one. Since every dual pair of vector spaces has the FARKAS Property, these operations appear to preserve this property, if they are applied to vector spaces. It seems to be natural therefore, to force them to preserve this property, if they are applied to oriented matroids. This is done in the following definition. In addition, it is natural and convenient to require one more condition, corresponding to the fact that vector spaces are symmetric with respect to the origin. Let us say that a set $\mathcal{S} \subseteq 2^{\pm E}$ is **symmetric with respect to the origin** or simply **symmetric** if $\mathcal{S} = -\mathcal{S}$.

Definition 5.8 *Let $E \neq \emptyset$ be finite and $\mathcal{F}, \mathcal{F}' \subseteq 2^{\pm E}$. Then $(\mathcal{F}, \mathcal{F}')$ is called an* **oriented matroid** *("OM") on E, if*

(a) $\mathcal{F}$ and $\mathcal{F}'$ are symmetric and

(b) every reorientation of every minor of $(\mathcal{F}, \mathcal{F}')$ has the FARKAS *Property.*

Obviously, the definition is symmetric, so $(\mathcal{F}, \mathcal{F}')$ is an OM if and only if $(\mathcal{F}', \mathcal{F})$ is one. Furthermore, every minor and every reorientation of an OM is an OM again, i.e. the class of oriented matroids is closed under taking minors and reorientations in the same way as the class of vector spaces is closed under these operations.

Example 5.9 *If $(L, L^\perp)$ is a dual pair of subspaces of $\mathbb{K}^E$, then the corresponding pair $(\mathcal{F}, \mathcal{F}')$ of sign vector sets is an OM on E.*

Example 5.10 *Suppose that $(\mathcal{H}, \mathcal{H}')$ and $(\mathcal{F}, \mathcal{F}')$ are OMs on E and $\mathcal{H} \subseteq \mathcal{F}$, $\mathcal{H}' \subseteq \mathcal{F}'$. Then every pair $(\mathcal{S}, \mathcal{S}')$, satisfying $\mathcal{H} \subseteq \mathcal{S} \subseteq \mathcal{F}$ and $\mathcal{H}' \subseteq \mathcal{S}' \subseteq \mathcal{F}'$ is an OM provided $\mathcal{S}$ and $\mathcal{S}'$ are symmetric. The proof is left to the reader as an exercise.*

Example 5.11 *Let $G = (V, E)$ be a digraph. Let $\mathcal{C}$ and $\mathcal{C}^*$ denote the sets of incidence vectors of circuits and cocircuits, resp. If we consider $\mathcal{C}$ and $\mathcal{C}^*$ as sets of sign vectors in the obvious way, then $(\mathcal{C}, \mathcal{C}^*)$ becomes an oriented matroid on E. This can be seen as follows. We know already, that $(\mathcal{C}, \mathcal{C}^*)$ has the* FARKAS *Property (cf. Example 5.3). Let L denote the circuit space of G, thus $\mathcal{C} = elem\, L$ and $\mathcal{C}^* = elem\, L^\perp$. Now consider a minor $\mathcal{C} \setminus I/J$ of $\mathcal{C}$, and let $G \setminus I/J$ denote the corresponding minor of G. Then $L \setminus I/J$ is the circuit space of $G \setminus I/J$ (cf. Example 5.5) and*

$$elem(L \setminus I/J) \subseteq (elem\, L) \setminus I/J \subseteq L \setminus I/J.$$

(The first inclusion is a straightforward consequence of the definition of minors and elementary vectors.) Similarily, we get that

$$elem(L^\perp \setminus J/I) \subseteq (elem\, L^\perp) \setminus J/I \subseteq L^\perp \setminus J/I.$$

Thus $(\mathcal{C} \setminus I/J, \mathcal{C}^* \setminus J/I)$ is "in between" two pairs that do have the FARKAS *Property. Furthermore, it is obviously symmetric. Thus the claim follows from Example 5.10.*

Proposition 5.12 (MINTY's Lemma) *Let $\mathcal{F}$ and $\mathcal{F}'$ be sets of sign vectors in $2^{\pm E}$ which are symmetric. Then $(\mathcal{F}, \mathcal{F}')$ is an OM on E if and only if it has the* MINTY **Property**, *i.e.*

(MP) *For every partition $E = R \cup G \cup B \cup W$ into four sets (i.e. every element of E is coloured red, green, blue or white) and for every $e \in R \cup G$ either*

(a) $\exists U \in \mathcal{F}$, $e \in supp\, U$, $U_R \geq 0$, $U_G \leq 0$, $U_B = *$, $U_W = 0$

or

(b) $\exists Y \in \mathcal{F}'$, $e \in supp\, Y$, $Y_R \geq 0$, $Y_G \leq 0$, $Y_B = 0$, $Y_W = *$

but not both.

Proof. Let $(\mathcal{F}, \mathcal{F}')$ be a pair of sets of sign vectors on E and let $E = R \cup G \cup B \cup W$ be a partition. Let $(\mathcal{H}, \mathcal{H}') := ({}_{\bar{G}}\mathcal{F} \setminus W/B, {}_{\bar{G}}\mathcal{F}'/W \setminus B)$. Then $(\mathcal{H}, \mathcal{H}')$ has the FARKAS Property, if and only if for every $e \in R \cup G$ either

(a) $\exists U \in \mathcal{H},\ e \in supp\, U,\ U \geq 0$

or

(b) $\exists Y \in \mathcal{H}',\ e \in supp\, Y,\ Y \geq 0$

but not both.

By the definition of minors, this means that either

(a) $\exists U \in {}_{G}\mathcal{F},\ e \in supp\, U,\ U_{R\cup G} \geq 0,\ U_B = *,\ U_W = 0$

or

(b) $\exists Y \in {}_{G}\mathcal{F}',\ e \in supp\, Y,\ Y_{R\cup G} \geq 0,\ Y_B = 0,\ Y_W = *$

but not both.

This is equivalent to the condition in the claim.

□

Remark 5.13 MINTY*'s Lemma may be visualized as indicated in the note following Definition 5.6: If $E = R \cup G \cup B \cup W$ is a partition of E and, say, $e \in R$, then either*

$$\overbrace{e\qquad\qquad}^{R}\overbrace{\qquad\qquad}^{G}\overbrace{\qquad\qquad}^{B}\overbrace{\qquad\qquad}^{W}$$

$$\exists X = +\oplus\cdots\oplus\ominus\cdots\ominus\ *\cdots*\ 0\cdots0 \in \mathcal{F}$$

or

$$\exists Y = +\oplus\cdots\oplus\ominus\cdots\ominus\ 0\cdots0\ *\cdots*\ \in \mathcal{F}'$$

but not both.

Here, clearly, "$\oplus$" stand for "+" or "0", "$\ominus$" stands for "−" or "0" and "$*$" stands for "+" or "−" or "0".

At the first glance, Proposition 5.12 seems quite "technical". (Indeed, it will turn out to be a powerful technical tool.) We will see later how MINTY's Lemma can be used, to prove the "Max flow- min cut"-Theorem. Then, each of the four colours will have a special meaning and the theorem may be understood more clearly.

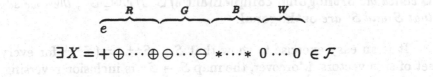

5.4 Abstract Orthogonality

Recall that our original problem (cf. the introduction to Chapter 5) has been the following: Given an *OM* $(\mathcal{F}, \mathcal{F}')$, does there exist a corresponding pair of orthogonal subspaces $(L, L^{\perp})$? Here we will give an answer to a related (but simpler) question, namely: Given an *OM* $(\mathcal{F}, \mathcal{F}')$, does there exist a kind of orthogonality relation between $\mathcal{F}$ and $\mathcal{F}'$?

If $x, y \in \mathbb{K}^E$, then x and y are orthogonal, if their inner product $x^T y$ equals zero. Since there is no way to "add" signs, there seems to be no reasonable way to mimic this definition for sign vectors. So we have to try to get on with a weaker one. Note that if $x, y \in \mathbb{K}^E$ "agree in sign" in some coordinate $e \in E$ (e.g. $x_e > 0$ and $y_e > 0$), then $x \perp y$ implies that x and y "differ in sign" in some other coordinate $f \in E$. This motivates the following definition:

Definition 5.14 *Two sign vectors $X, Y \in 2^{\pm E}$ are called* **orthogonal**, *if*

$$(X^+ \cap Y^+) \cup (X^- \cap Y^-) \neq \emptyset \Leftrightarrow (X^+ \cap Y^-) \cup (X^- \cap Y^+) \neq \emptyset.$$

This is denoted by $X \perp Y$.

If $\mathcal{S} \subseteq 2^{\pm E}$, then $X \in 2^{\pm E}$ is called **orthogonal** *to $\mathcal{S}$, denoted by $X \perp \mathcal{S}$, provided $X \perp Y$ for every $Y \in \mathcal{S}$. The set*

$$\mathcal{S}^{\perp} = \{X \in 2^{\pm E} \mid X \perp \mathcal{S}\}$$

is called the **orthogonal complement** *of $\mathcal{S}$. If $\mathcal{S}' \subseteq \mathcal{S}^{\perp}$, then we say that $\mathcal{S}$ and $\mathcal{S}'$ are* **orthogonal**.

It is an easy exercise to show that $\mathcal{S} \subseteq \mathcal{S}^{\perp\perp} := (\mathcal{S}^{\perp})^{\perp}$ for every set of sign vectors. Moreover, the map $\mathcal{S} \to \mathcal{S}^{\perp}$ is inclusion reversing, i.e. $\mathcal{S} \subseteq \mathcal{T}$ implies $\mathcal{S}^{\perp} \supseteq \mathcal{T}^{\perp}$.

Example 5.15 *Let $\mathcal{C}$ and $\mathcal{C}^*$ denote the sets of incidence vectors of circuits and cocircuits, resp., of a digraph $G = (V, E)$. Let $x \in \mathcal{C}$ and $y \in \mathcal{C}^*$, and let X and Y denote the corresponding sign vectors. Then $X \perp Y$ expresses the fact that if the circuit x passes the cut y in one direction, it must also pass it in the opposite direction.*

Example 5.16 *Let $L \leq \mathbb{K}^E$ be a subspace. Since, as we noted already, "abstract orthogonality" between sign vectors is* **weaker** *than "ordinary orthogonality" between vectors, it follows that*

$$\sigma(L)^{\perp} \supseteq \sigma(L^{\perp}).$$

We will show later that actually equality holds, indicating that the abstract version of orthogonality is (still) quite reasonable. The reader is invited to check this in the special case of Example 5.2. Here we will content ourselves by showing that orthogonality between the two "parts" $\mathcal{F}$ and $\mathcal{F}'$ of an OM $(\mathcal{F}, \mathcal{F}')$ is an easy consequence of MINTY's Lemma:

Theorem 5.17 *Let $\mathcal{F}, \mathcal{F}' \subseteq 2^{\pm E}$ be two sets of sign vectors. Then $\mathcal{F}$ and $\mathcal{F}'$ are orthogonal if and only if for every partition $E = R \cup G \cup B \cup W$ and every $e \in R \cup G$ at most one of the alternatives (a) and (b) in (MP) holds.*

Proof. Let $E = R \cup G \cup B \cup W$ be a partition of E, and let $e \in E$. If there exists $X \in \mathcal{F}$ and $Y \in \mathcal{F}'$ as in a) and b) of Proposition 5.12, then $e \in (X^+ \cap Y^+) \cup (X^- \cap Y^-)$, while $(X^+ \cap Y^-) \cup (X^- \cap Y^+) = \emptyset$. Thus $\mathcal{F}$ and $\mathcal{F}'$ are not orthogonal. Conversely, suppose that $\mathcal{F}$ and $\mathcal{F}'$ are not orthogonal. Let $X \in \mathcal{F}$ and $Y \in \mathcal{F}'$ such that X and Y are not orthogonal. Assume that, say, $(X^+ \cap Y^+) \cup (X^- \cap Y^-) \neq \emptyset$ and $(X^+ \cap Y^-) \cup (X^- \cap Y^+) = \emptyset$. Now let $R := X^+ \cap Y^+$, $G := X^- \cap Y^-$, $B := Y^0$, $W := X^0 \setminus Y^0$ and $e \in R \cup G$. Then condition (a) of Proposition 5.12 is satisfied as well as (b).

□

We obtain at once two corollaries:

Corollary 5.18 *If $(\mathcal{F}, \mathcal{F}')$ is an OM, then $\mathcal{F}$ and $\mathcal{F}'$ are orthogonal, i.e. $\mathcal{F}' \subseteq \mathcal{F}^\perp$.*

Corollary 5.19 *If $(\mathcal{F}, \mathcal{F}')$ is an OM, so is $(\mathcal{F}, \mathcal{F}^\perp)$.*

Proof. If $(\mathcal{F}, \mathcal{F}')$ is an OM, then $\mathcal{F}' \subseteq \mathcal{F}^\perp$. Since $(\mathcal{F}, \mathcal{F}')$ has the MINTY Property, $\mathcal{F}' \subseteq \mathcal{F}^\perp$ implies that for the pair $(\mathcal{F}, \mathcal{F}^\perp)$ at least one of the alternatives (a) and (b) in (MP) holds. On the other hand, Theorem 5.17 shows that at most one of them can hold.

□

5.5 Abstract Elimination Property

In linear algebra, one usually considers **single** linear vector spaces rather than pairs of orthogonal vector spaces $(L, L^\perp)$. There one starts with a set L of "vectors" (and some field F) and defines L to be a vector space, if certain axioms are satisfied. In the sequel we will try to proceed in a similar way: We will start with a set of sign vectors and define it to be an "oriented matroid", if it has certain properties.

Since the essential axiom for vector spaces is, that they are closed with respect to addition of vectors — an operation which doesn't make sense for sign vectors —, we have to look for another property of vector spaces which is more "combinatorial" in nature, allowing to interprete it in the context of sign vectors. This property will get the name "elimination property". Again, let us first consider some examples in order to see what this property is like.

Example 5.20 *Let $L \leq \mathbb{K}^E$ be a subspace, and let $x, y \in L$. We say that $e \in E$* **separates** *x and y, if x and y have opposite signs in coordinate e. Equivalently, x and y lie on opposite sides of the hyperplane $H_e := \{w \in \mathbb{K}^n \mid w_e = 0\}$, i.e. H_e separates x and y. If e separates x and y, then the line segment $[x, y]$ intersects H_e in a point (vector) $z = \lambda x + \mu y$, $\lambda > 0$, $\mu > 0$ and $\lambda + \mu = 1$. The e-th coordinate of z is equal to zero, and therefore, we will say that z arises by* **eliminating** e **between** x **and** y . *If in addition $x_f \neq 0$ for some $f \in E$ which does not separate x and y, then (since $\lambda > 0$ and $\mu > 0$) the f-th coordinate of z has the same sign as the f-th coordinate of x. We therefore say, that z arises by* **eliminating** e **between** x **and** y, **fixing** f. .

Example 5.21 *Let $G = (V, E)$ be a digraph, let L denote its circuit space, and let $\mathcal{F}$ denote the corresponding set of sign vectors.*

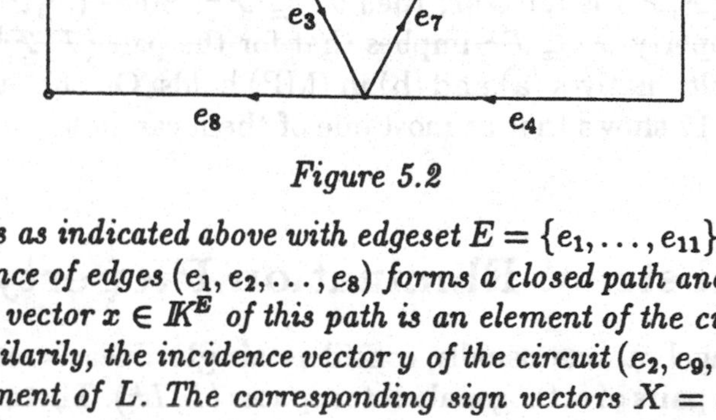

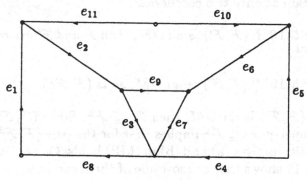

Figure 5.2

If G is as indicated above with edgeset $E = \{e_1, \ldots, e_{11}\}$, then, e.g. the sequence of edges $(e_1, e_2, \ldots, e_8)$ forms a closed path and hence the incidence vector $x \in \mathbb{K}^E$ of this path is an element of the circuit space of G. Similarily, the incidence vector y of the circuit $(e_2, e_9, e_6, e_{10}, e_{11})$ is an element of L. The corresponding sign vectors $X = \sigma(x)$, $Y = \sigma(y)$ in $\mathcal{F}$ are, resp.:

$$\begin{aligned} X &= (+,+,+,-,+,+,-,+,0,0,0) \\ Y &= (0,+,0,0,0,-,0,0,+,-,+) \end{aligned}$$

Obviously, X and Y are separated by e_6 and, e.g. e_1 is an element of $\operatorname{supp} X$ which does not separate X and Y. Then for example, the sign vectors

$$\begin{aligned} Z &= (+,+,+,0,0,0,0,+,0,0,0) \in \mathcal{F} \\ \text{and } Z' &= (+,+,0,0,0,0,-,+,+,0,0) \in \mathcal{F} \end{aligned}$$

corresponding to the circuits (e_1, e_2, e_3, e_8) and $(e_1, e_2, e_9, e_7, e_8)$ are obtained by eliminating e_6 between X and Y, fixing e_1.

Definition 5.22 *Let $X, Y \in 2^{\pm E}$. Then $e \in E$* **separates** *X and Y if X_e and Y_e are opposite signs. The set of elements $e \in E$ separating X and Y is denoted by* **sep**(X, Y). *Suppose that $e, f \in \operatorname{supp} X \cup \operatorname{supp} Y$ such that e separates X and Y, but f does not. Then $Z \in 2^{\pm E}$ is said to be obtained by* **eliminating e between X and Y, fixing f,** *if*

$$e \notin \operatorname{supp} Z,\ f \in \operatorname{supp} Z,\ Z^+ \subseteq X^+ \cup Y^+ \text{ and } Z^- \subseteq X^- \cup Y^-.$$

The set $\mathcal{S} \subseteq 2^{\pm E}$ is said to have the **elimination property,** *if the following holds:*

(EP) *If $X, Y \in \mathcal{S}$, $e, f \in \operatorname{supp} X \cup \operatorname{supp} Y$ such that e separates X and Y, but f does not, then there exists a $Z \in \mathcal{S}$ which is obtained by eliminating e between X and Y, fixing f.*

(Note that if X and Y are separated by some $c \in E$, then there always exists a nonseparating element f, unless $X = -Y$.)

Theorem 5.23 *Let $(\mathcal{F}, \mathcal{F}')$ be an OM on E. Then $\mathcal{F}$ has the elimination property.*

Proof. Let $X, Y \in \mathcal{F}$ and suppose that e and f are as above. To apply MINTY's Lemma, let

$$\begin{aligned} R &:= (X^+ \cup Y^+) \setminus (X^- \cup Y^-), \\ G &:= (X^- \cup Y^-) \setminus (X^+ \cup Y^+), \\ B &:= [(X^+ \cap Y^-) \cup (X^- \cap Y^+)] \setminus e, \\ W &:= X^0 \cap Y^0 \cup e. \end{aligned}$$

Thus, in particular, $f \in R \cup G$, and hence either

a) $\exists Z \in \mathcal{F},\ f \in \operatorname{supp} Z,\ Z_R \geq 0,\ Z_G \leq 0,\ Z_B = *,\ Z_W = 0$

or

b) $\exists Z' \in \mathcal{F}'$, $f \in supp\, Z'$, $Z'_R \geq 0$, $Z'_G \leq 0$, $Z'_B = 0$, $Z'_W = *$.

Obviously, if a) holds, we are done. Thus let us show, that b) can not hold. Suppose $Z' \in \mathcal{F}'$ is as in b). Recall that, by assumption, $f \in supp\, X \cup supp\, Y$. Suppose w.l.o.g. that $f \in supp\, X$, say $f \in X^+$. This may be illustrated as follows:

$$\begin{array}{rcll} \exists X & = & +\oplus\cdots\oplus\ominus\cdots\ominus*\cdots*\,0\cdots0\,* & \in \mathcal{F} \\ \exists Y & = & \oplus\ \ \cdots\oplus\ominus\cdots\ominus*\cdots*\,0\cdots0\,* & \in \mathcal{F} \\ \exists Z' & = & \underbrace{\overset{f}{+}\oplus\cdots\oplus}_{R}\underbrace{\ominus\cdots\ominus}_{G}\underbrace{0\cdots0}_{B}\underbrace{*\cdots*\overset{e}{*}}_{W} & \in \mathcal{F}' \end{array}$$

Since $X, Y \in \mathcal{F}$ and $Z' \in \mathcal{F}'$, we conclude that $X \perp Z'$ and $Y \perp Z'$. $X \perp Z'$ implies that $Z'_e = -X_e$ (since $Z'_f = X_f$ and X and Z' cannot have opposite signs in any coordinate different from e). Since X and Y are separated by e, this implies $Z'_e = Y_e$, showing that Z' and Y are not orthogonal, a contradiction.

□

We will see later, that the elimination property (EP) can be used to derive an alternative definition of oriented matroids. However, before we can do that, we first have to understand more clearly, why $\mathcal{F}$ has this property, if $(\mathcal{F}, \mathcal{F}')$ is an oriented matroid. In the following, we will define elementary sign vectors in analogy to the definition of elementary vectors of vector spaces and investigate the structure of OMs more closely.

5.6 Elementary vectors

Recall that an elementary vector of a vector space L is a nonzero element of L with minimal support. This definition carries over very naturally to sign vectors:

Definition 5.24 *Let $\mathcal{S} \subseteq 2^{\pm E}$ be a set of sign vectors. Then a nonzero $X \in \mathcal{S}$ is called an* **elementary** *sign vector of $\mathcal{S}$, if X has minimal nonempty support, i.e. if $\emptyset \neq supp\, Y \subseteq supp\, X$ implies $supp\, Y = supp\, X$ for every $Y \in \mathcal{S}$. The set of all elementary sign vectors in $\mathcal{S}$ is denoted by* **elem** $\mathcal{S}$.

Example 5.25 *Let $L \leq K^E$ be a subspace and $\mathcal{F} := \sigma(L)$. Then $elem\, \mathcal{F} = \sigma(elem\, L)$. In particular, if L denotes the circuit space of a digraph G, there is a $1-1$ correspondence between elements of $elem\, \mathcal{F}$ and incidence vectors of circuits of G. If C is a circuit of G and v is* *any vertex incident to* C*, then there exist exactly two edges* e, f *of* C *which are incident to* v*. Thus, if* $X \in \mathrm{elem}\,\mathcal{F}$ *denotes the sign vector corresponding to* C *and* Y *denotes the sign vector corresponding to the cocircuit consisting of all edges incident to* v*, then* $\mathrm{supp}\,X \cap \mathrm{supp}\,Y = \{e, f\}$*. We will see below that this property characterizes elementary vectors.*

Example 5.26 *More generally, let* $L = \ker A \leq \mathbb{K}^E$ *be a subspace defined by some matrix* A *with columns indexed by* E*. Let* $\mathcal{F} = \sigma(L)$*, and let* $X \in \mathcal{F}$*. Then* $\mathrm{supp}\,X$ *corresponds to a (linearly) dependent set* $C = \{A_{.e} \mid e \in \mathrm{supp}\,X\}$ *of columns of* A*. Hence* X *is elementary if and only if* C *is a minimal dependent subset of the columns of* A*. This means that every proper subset of* C *is independent. In particular, if* $A_{.e}, A_{.f} \in C$*, then*

$$M := \mathrm{lin}(C \setminus \{A_{.e}, A_{.f}\}) < \mathrm{lin}\,C := N \leq \mathbb{K}^E.$$

Hence there exists $u \in M^\perp \setminus N^\perp$*. Let* $y = u^T A \in L^\perp$*, and let* Y *denote the corresponding sign vector. Then, by definition of* u*,* $Y \in \mathcal{F}' := \sigma(L^\perp)$ *is such that* $\mathrm{supp}\,X \cap \mathrm{supp}\,Y = \{e, f\}$*.*

The most general version of this observation is the following:

Proposition 5.27 *Let* $(\mathcal{F}, \mathcal{F}')$ *be an OM on some finite set* E*, and let* $X \in \mathcal{F}$*. Then* $X \in \mathrm{elem}\,\mathcal{F}$ *iff for every two elements* $e, f \in \mathrm{supp}\,X$ *there exists a* $Y \in \mathcal{F}'$ *such that* $\mathrm{supp}\,Y \cap \mathrm{supp}\,X = \{e, f\}$*.*

Proof. Let $X \in \mathcal{F}$ and $\{e, f\} \subseteq \mathrm{supp}\,X$. We set $R := \{e\}$, $G := \emptyset$, $B := \mathrm{supp}\,X \setminus \{e, f\}$ and $W := X^0 \cup \{f\}$. Then MINTY's Lemma states that either

a) $\exists U \in \mathcal{F},\ e \in \mathrm{supp}\,U \subseteq \mathrm{supp}\,X \setminus \{f\}$

or

b) $\exists Y \in \mathcal{F}',\ e \in \mathrm{supp}\,Y \subseteq X^0 \cup \{e, f\}$

but not both.

Clearly, X is elementary if and only if a) is false for every set $\{e, f\} \subseteq \mathrm{supp}\,X$. Hence X is elemetary if and only if b) holds for every $\{e, f\} \subseteq \mathrm{supp}\,X$. On the other hand, if Y is as in b), then we must have $\{e, f\} \subseteq \mathrm{supp}\,Y$, since otherwise Y cannot be orthogonal to X. Hence, b) is satiesfied for some $\{e, f\} \subseteq \mathrm{supp}\,X$, if and only if

there exists $Y \in \mathcal{F}'$ such that $supp\, Y \cap supp\, X = \{e, f\}$. This proves the proposition.

□

Using the fact that $\mathcal{F}$ has the elimination property, we will now derive another characterization of the set $elem\,\mathcal{F}$. Note that, for a set $\mathcal{S}$ of sign vectors, we defined the elementary vectors of $\mathcal{S}$ to be those elements of $\mathcal{S}$ which are minimal with respect to the relation "$\leq$" on $\mathcal{S}$, defined by "$X \leq Y \Leftrightarrow supp\, X \subseteq supp\, Y$". One may ask, whether the partial order on $\mathcal{S}$, defined by this relation "$\leq$" is the most natural we can think of. Indeed, there is another canonical way to order the elements of $\mathcal{S}$:

Definition 5.28 *If $X, Y \in 2^{\pm E}$, we say that X* **conforms** *to Y, if $X^+ \subseteq Y^+$ and $X^- \subseteq Y^-$. This is denoted by $X \preceq Y$.*

Example 5.29 *Let G be the digraph with edge set $E = \{e_1, \ldots, e_7\}$ as indicated below, let L be its circuit space and $\mathcal{F} := \sigma(L)$.*

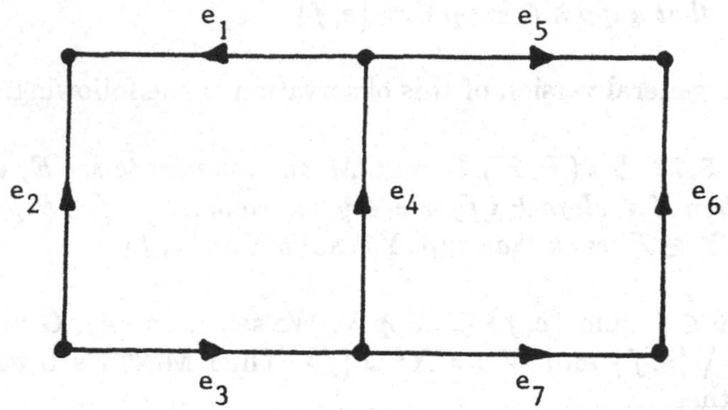

The vectors $x := (1, -1, 1, 1, 0, 0, 0)$ and $x' := (0, 0, 0, 1, 1, -1, -1)$ are incidence vectors of circuits of G and hence elements of L. Thus the vector $y := x + x' = (1, -1, 1, 2, 1, -1, -1,)$ is an element of L. The sign vectors X and X' of x and x' both conform to the sign vector Y of y.

If $\mathcal{S}$ is a set of sign vectors and $X \in elem\,\mathcal{S}$, then, by definition, X is minimal nonzero with respect to the relation "$\preceq$" on $\mathcal{S}$. If $\mathcal{S}$ happens to belong to an oriented matroid $(\mathcal{S}, \mathcal{S}')$, then the converse is also true:

Lemma 5.30 *If $\mathcal{F} \subseteq 2^{\pm E}$ has the elimination property, then elem $\mathcal{F}$ is the set of sign vectors which are minimal nonzero in $\mathcal{F}$ with respect to the relation "$\preceq$".*

Proof. Let $X_1 \in \mathcal{F}$ be minimal nonzero with respect to the relation "$\preceq$" on $\mathcal{F}$. We have to show that $X_1 \in elem\,\mathcal{F}$. Suppose, it is not, i.e. there exists a nonzero $X_2 \in \mathcal{F}$ such that $supp\,X_2 \subset supp\,X_1$. Since X_1 is minimal with respect to "$\preceq$", X_2 cannot conform to X_1. Let $e \in E$ separate X_1 and X_2, and let $f \in supp\,X_1 \setminus supp\,X_2$ (thus f does not separate X_1 and X_2). Eliminating e between X_1 and X_2, fixing f we obtain a nonzero $X_3 \in \mathcal{F}$ such that $supp\,X_3 \subset supp\,X_1$ and $X_3^+ \subseteq (X_1^+ \cup X_2^+) \setminus e$, $X_3^- \subseteq (X_1^- \cup X_2^-) \setminus e$. Thus the number of elements in E which separate X_1 and X_3 is strictly less than the number of elements which separate X_1 and X_2. Continuing this process, eliminating between X_1 and X_3, we will finally end up with some $X_k \in \mathcal{F}$ such that $supp\,X_k \subset supp\,X_1$ and the number of elements separating X_1 and X_k is equal to zero, i.e. X_k conforms to X_1. This contradicts the minimality of X_1 with respect to "$\preceq$".

□

5.7 The Composition Theorem

Our goal in this section is to show that every vector in $\mathcal{F}$ is in a sense "composed" of elementary sign vectors. To get an idea of what we mean by this, consider the digraph G as in Example 5.29. The two sign vectors $X = (+,-,+,+,0,0,0)$ and $X' = (0,0,0,+,+,-,-)$ are both elementary sign vectors in $\mathcal{F}$ which are not separated by any element $e \in E$. Their "sum" yields the sign vector $Y = (+,-,+,+,+,-,-) \in \mathcal{F}$. Our aim is to show that in general, every sign vector in $\mathcal{F}$ is the "sum" of pairwise "unseparated" elementary sign vectors.

Definition 5.31 *Let $X, Y \in 2^{\pm E}$ be two sign vectors. Then the sign vector $Z \in 2^{\pm E}$, defined by $Z_e := X_e$ if $X_e \neq 0$ and $Z_e = Y_e$ otherwise, is called the* **composition** *of X and Y, and is denoted by $X \circ Y$. Two sign vectors X and Y are called* **compatible**, *if $sep(X, Y) = \emptyset$. Thus $X \circ Y = Y \circ X$ if and only if X and Y are compatible and $X \circ Y = Y$ if and only if X conforms to Y. Note that "$\circ$" is associative, hence we may ommit brackets in expressions like $X_1 \circ X_2 \circ \ldots \circ X_k$. If $\mathcal{S} \subseteq 2^{\pm E}$, then*

$$span\,\mathcal{S} := \{X_1 \circ \ldots \circ X_k \mid X_1, \ldots, X_k \in \mathcal{S}, k \geq 0\}.$$

We say that $Y = X_1 \circ \ldots \circ X_k$ is the **conformal sum** *of $X_1, \ldots, X_k$, if each of $X_1, \ldots, X_k$ conforms to Y, i.e. if $X_1, \ldots, X_k$*

are pairwise compatible. By convention, the zero vector is the sum over the empty set.

Example 5.32 *Let $L \leq \mathbb{K}^E$ be a vector space, and let $\mathcal{F} := \sigma(L)$. Given $x, y \in L$, we say that x conforms to y, if the sign vector $X = \sigma(x)$ conforms to $Y = \sigma(y)$. Similarily, x and y are called compatible, if their corresponding sign vectors are compatible. A vector $z \in L$ is called the composition of x and y, if the corresponding is true for the sign vectors of x, y and z. Finally, if $x_1, \ldots, x_k \in L$, $y \in L$ is called the conformal sum of the x_i's, if $y = x_1 + \ldots + x_k$ and the x_i's all conform to y.*

Given $x, y \in L$, it is easy to determine $z \in L$ such that z is the composition of x and y: If we choose $\lambda > 0$ large enough, then $z := \lambda x + y$ will have the desired properties.

Lemma 5.33 *Let $\mathcal{S}$ and $\mathcal{S}'$ be orthogonal sets of sign vectors. Then $span\,\mathcal{S}$ and $span\,\mathcal{S}'$ are orthogonal, too.*

Proof. It is easy to see, that for $X, Y, U \in 2^{\pm E}$, $X \perp U$ and $Y \perp U$ imply $X \circ Y \perp U$. From this the claim follows by induction.

□

Proposition 5.34 *Let $(\mathcal{F}, \mathcal{F}')$ be an OM. Then $(\mathcal{F}, span\,\mathcal{F}')$, and hence $(span\,\mathcal{F}, \mathcal{F}')$ and $(span\,\mathcal{F}, span\,\mathcal{F}')$ are OMs, too.*

Proof. Since $\mathcal{F}' \subseteq span\,\mathcal{F}' \subseteq \mathcal{F}^\perp$ by Lemma 5.33 and $(\mathcal{F}, \mathcal{F}^\perp)$ is an OM by Corollary 5.19, the result follows (cf. Example 5.10).

□

As we noted already, we want to show that every element of $\mathcal{F}$ is the conformal sum of elementary sign vectors. In case, $\mathcal{F}$ arises from a vector space L, the proof is easy:

Proposition 5.35 *Let $L \leq \mathbb{K}^E$. Then every vector in L is the conformal sum of elementary vectors.*

Proof. Let $y \in L$. The proof will be by induction on $s := |supp(y)|$. If $s = 0$, then $y = 0$ is the empty sum. Thus let $s \geq 1$ and assume, the claim holds for every vector in L which has less than s elements in its support. By Lemma 5.30, there exists $x \in elem\,L$ such that $x \preceq y$. It is easy to see, that we can find $\lambda > 0$ such that $y' := y - \lambda x$ (still) conforms to y and $|supp\,y'| < s$. Thus, applying the induction hypothesis to y', it follows that $y' = x_1 + \ldots + x_k$ with $x_1, \ldots, x_k \in elem\,L$ conforming to y'. Since $x_{k+1} := \lambda x$ is elementary, too, we get that $y = x_1 + \ldots + x_{k+1}$ is the conformal sum of elementary vectors.

□

Note, that in the proof of Proposition 5.35, we used some scaling techniques, which cannot be applied to sign vectors. This is one of the (very rare) situations in which a proof technique does not carry over into the context of sign vectors. Nevertheless, the result remains true also in the general setting:

Theorem 5.36 (Composition Theorem) *If $\mathcal{F} \subseteq 2^{\pm E}$ is symmetric and has the elimination property, then every element $Y \in span\,\mathcal{F}$ (and hence every $Y \in \mathcal{F}$) is the conformal sum of elementary sign vectors in $\mathcal{F}$. In particular, this is true, if $(\mathcal{F}, \mathcal{F}')$ is an OM on E.*

Proof. We will first show that every element $Y \in \mathcal{F}$ is such a conformal sum. The proof is by induction on $s := |supp\,Y|$. If $s = 1$, then Y is obviously elementary and there is nothing to prove in this case. Suppose now that $s > 1$ and that the claim holds for every element of $\mathcal{F}$ which has less then s elements in its support. Let $Y \in \mathcal{F}$ and $|supp\,Y| = s$. We will show that for every $f \in supp\,Y$ there exists an element $X \in elem\,\mathcal{F}$ which conforms to Y and has f in its support. Then, clearly, Y is the conformal sum of all these elementary sign vectors X, and we are done.

Thus let $f \in supp\,Y$. Let $X \in elem\,\mathcal{F}$ such that X conforms to Y. (Note that existence is guaranteed by Lemma 5.30). If $f \in supp\,X$, we are done. Thus, suppose that $f \notin supp\,X$. Let $Y_1 := -X$ and choose some element $e \in sep(Y, Y_1)$. Now we proceed in the same way as in the proof of Lemma 5.30: We eliminate e between Y and Y_1, fixing f, to get some $Y_2 \in \mathcal{F}$ with $f \subset supp\,Y_2 \subset supp\,Y$. If Y_2 does not conform to Y, we continue the elimination process, eliminating some element between Y and Y_2, fixing f. This yields some $Y_3 \in \mathcal{F}$ with $f \in supp\,Y_3 \subset supp\,Y$ and the number of elements which separate Y and Y_3 is strictly less than the number of elements separating Y and Y_2. Thus after at most $s-1$ steps, we will end up with some $Y_k \in \mathcal{F}$ such that $f \in supp\,Y_k \subset supp\,Y$ and $Y_k \preceq Y$. Since $|supp\,Y_k| < s$, our induction hypothesis applies, i.e. Y_k is the conformal sum of elementary vectors in $\mathcal{F}$. In particular, there exists $X' \in elem\,\mathcal{F}$ such that $f \in supp\,X'$ and $X' \preceq Y_k \preceq Y$, and we are done.

Now let $Y = Y_1 \circ \ldots \circ Y_k \in span\,\mathcal{F}$, $Y_1, \ldots, Y_k \in \mathcal{F}$. If $Y_1, \ldots, Y_k$ are pairwise compatible, then Y is the conformal sum of elementary vectors, since each of $Y_1, \ldots, Y_k$ is. We proceed by induction on the number t of elements $e \in E$ which separate any two of the sign vectors $Y_1, \ldots, Y_k$. If $t \geq 1$, we will show that we can write Y as $Y = Y_1' \circ \ldots \circ Y_l'$ for some $Y_1', \ldots, Y_l' \in \mathcal{F}$ such that the number of elements separating some of the sign vectors $Y_1', \ldots, Y_l'$ is strictly less than t. Thus, let $e \in E$ separate Y_i and Y_j, $i \leq j$. For every $f \in Y_i^\circ \cap supp\,Y_j$, let

$Y(f) \in \mathcal{F}$ be obtained by eliminating e between Y_i and Y_j, fixing f. Then we replace Y_j in $Y_1 \circ \ldots \circ Y_k$ by the composition of all these $Y(f)$'s to get $Y = Y_1' \circ \ldots \circ Y_n'$. It is easy to see that the number of elements separating any two of $Y_1', \ldots, Y_n'$ is $\leq t$. Thus, if we proceed eliminating e between every pair Y_i and Y_j such that e separates Y_i and Y_j, we will finally end up with a representation $Y = Y_1' \circ \ldots \circ Y_l'$ such that e does no longer separate any two of $Y_1', \ldots, Y_l'$.

□

There are many consequences of the Composition Theorem. We just list some of them below:

Corollary 5.37 *Let $(\mathcal{F}, \mathcal{F}')$ be an OM on E. Then $(elem\,\mathcal{F})^\perp = \mathcal{F}^\perp$.*

Proof. $\mathcal{F}^\perp \subseteq (elem\,\mathcal{F})^\perp$ is trivial. To prove the converse inclusion, let $U \in (elem\,\mathcal{F})^\perp$, i.e. $U \perp X$ for every $X \in elem\,\mathcal{F}$. It is easily seen that this implies $U \perp Y$ for every conformal sum $Y = X_1 \circ \ldots \circ X_k$ of elementary vectors $X_i \in elem\,\mathcal{F}$. Hence $U \in \mathcal{F}^\perp$.

□

Corollary 5.38 *Let $(\mathcal{F}, \mathcal{F}')$ be an OM on some set E. Then $(\mathcal{F}, elem\,\mathcal{F}')$, and hence $(elem\,\mathcal{F}, \mathcal{F}')$ and $(elem\,\mathcal{F}, elem\,\mathcal{F}')$ are OMs, too.*

Proof. We will verify MINTY's Property for $(\mathcal{F}, elem\,\mathcal{F}')$ (cf. Proposition 5.12). Let $E = R \cup G \cup B \cup W$ be a partition of E, and let $e \in R \cup G$. Consider the two alternatives

a) $\exists U \in \mathcal{F}, \quad e \in supp\,U,\ U_R \geq 0,\ U_G \leq 0,\ U_B = *,\ U_W = 0$

and

b) $\exists Y \in elem\,\mathcal{F}',\ e \in supp\,Y,\ Y_R \geq 0,\ Y_G \leq 0,\ Y_B = 0,\ Y_W = *$.

Since $elem\,\mathcal{F}' \subseteq \mathcal{F}'$, at most one of the alternatives a) and b) can hold. Thus we have to show that **at least** one of a) and b) holds. This, however, is an easy consequence of the composition theorem.

□

Corollary 5.39 *Let $(\mathcal{F}, \mathcal{F}')$ be an OM. Then $elem\,\mathcal{F}$ has the elimination property.*

Proof. Corollary 5.38 and Theorem 5.23.

□

Perhaps the meaning of this result is most clearly understood, if one considers a digraph G: Given two circuits U_1 and U_2 in G and an edge e such that U_1 and U_2 "pass through e" in different directions, then we can find a circuit U in G that does not pass through e, but has the same direction as U_1 or U_2 in every edge that is not passed in different directions by U_1 and U_2. (cf. Example 5.21)

Given an OM $(\mathcal{F}, \mathcal{F}')$ then by Corollary 5.38, $(elem\,\mathcal{F}, elem\,\mathcal{F}')$ is an OM and by Proposition 5.34, $(span\,\mathcal{F}, span\,\mathcal{F}')$ is an OM, too. In the following, we will show that these are actually the "minimal" and "maximal" OMs which are contained in, resp. contain $(\mathcal{F}, \mathcal{F}')$. This will imply that $\sigma(L^\perp) = \sigma(L)^\perp$ for every vector space $L \leq \mathbb{K}^E$.

Proposition 5.40 *Let $(\mathcal{F}, \mathcal{F}')$ and $(\mathcal{H}, \mathcal{F}')$ be two OMs on E. Then $span\,\mathcal{F} = span\,\mathcal{H}$ and $elem\,\mathcal{F} = elem\,\mathcal{H}$.*

Proof. By Corollary 5.38, $(elem\,\mathcal{F}, elem\,\mathcal{F}')$ and $(elem\,\mathcal{H}, elem\,\mathcal{F}')$ are both OMs on E.
Let $X \in elem\,\mathcal{F}$. Let $R := X^+$, $G := X^-$, $B := \emptyset$ and $W := X^0$. Then the following holds for every $e \in R \cup G = supp\,X$:

$$\exists Z \in elem\,\mathcal{F}, e \in supp\,Z, Z_R \geq 0, Z_G \leq 0, X_W = 0.$$

Hence,

$$\nexists Y \in elem\,\mathcal{F}', e \in supp\,Y, Y_R \geq 0, Y_G \leq 0, Y_W = *.$$

Thus, for every $e \in supp\,X$

$$\exists Z(e) \in elem\,\mathcal{H}, e \in supp\,Z(e), Z(e)_R \geq 0, Z(e)_G \leq 0, Z(e)_W = 0.$$

The composition of all these $Z(e)'s$, $e \in supp\,X$, is equal to X. Hence, $X \in span(elem\,\mathcal{H}) = span\,\mathcal{H}$. Since X has been an arbitrary element of $elem\,\mathcal{F}$, this shows that $elem\,\mathcal{F} \subseteq span\,\mathcal{H}$, and consequently, using the composition theorem, $span\,\mathcal{F} = span(elem\,\mathcal{F}) \subseteq span\,\mathcal{H}$. By symmetry, $span\,\mathcal{F} = span\,\mathcal{H}$. This in turn gives $elem\,\mathcal{F} = elem(span\,\mathcal{F}) = elem(span\,\mathcal{H}) = elem\,\mathcal{H}$.

□

Corollary 5.41 *Let $(\mathcal{F}, \mathcal{F}')$ be an OM. Then $span\,\mathcal{F}' = \mathcal{F}^\perp$.*

Proof. By Corollary 5.19, $(\mathcal{F}, \mathcal{F}^\perp)$ is an OM. Furthermore, we conclude from Proposition 5.40, that $span\,\mathcal{F}' = span\,\mathcal{F}^\perp$. But $\mathcal{F}^\perp$ is obviously closed under conformal sums, i.e. $span\,\mathcal{F}^\perp = \mathcal{F}^\perp$.

□

Corollary 5.42 *If $L \leq \mathbb{K}^E$ is a vector space, then $\sigma(L^\perp) = \sigma(L)^\perp$.*

Proof. Since $(\sigma(L), \sigma(L^\perp))$ is an OM, it follows from Corollary 5.41, that $span(\sigma(L^\perp)) = \sigma(L)^\perp$. However, since $L^\perp$ is a vector space, $\sigma(L^\perp)$ is obviously closed under composition, i.e. $span(\sigma(L^\perp)) = \sigma(L^\perp)$.

□

5.8 Elimination Axioms

Recall that our aim has been to derive an alternative definition of oriented matroids by considering only a single set $\mathcal{F} \subseteq 2^{\pm E}$ rather than a dual pair $(\mathcal{F}, \mathcal{F}')$ of systems of sign vectors. Thus our goal is to characterize those sets $\mathcal{F} \subseteq 2^{\pm E}$ for which there exists a dual $\mathcal{F}' \subseteq 2^{\pm E}$ such that $(\mathcal{F}, \mathcal{F}')$ is an OM. The results of the last section imply that it suffices to characterize the sets *elem* $\mathcal{F}$ resp. *span* $\mathcal{F}$, where $(\mathcal{F}, \mathcal{F}')$ is an OM. Indeed, this has been the first approach to oriented matroids in the literature.

We start with defining oriented matroids in terms of their elementary sign vectors.

Definition 5.43 *Let $\emptyset \neq E$ be finite and $\mathcal{C} \subseteq 2^{\pm E}$. Then $\mathcal{C}$ is called an* **elementary oriented matroid** *("OM"), if the following holds:*

a) *$0 \notin \mathcal{C}$ and $\mathcal{C}$ is symmetric,*

b) *$\mathcal{C} = elem\, \mathcal{C}$,*

c) *$\mathcal{C}$ has the elimination property.*

Proposition 5.44 *If $(\mathcal{F}, \mathcal{F}')$ is an OM, then elem $\mathcal{F}$ is an elementary OM.*

Proof. Corollary 5.39. □

We refer to previous examples for an interpretation of the elimination Axiom c). In particular, the reader may try to find an "elementary" proof for the fact that the sign vector corresponding to the elementary vectors of some subspace $L \leq \mathbb{K}^E$ form an elementary OM. This is not completely trivial (once you agree that "elementary" means "without using the Composition Theorem"). Thus, so far, we have given two definitions of oriented matroids (which are not yet proved to be equivalent), and neither of them is trivially satisfied by vector spaces (resp. their elementary vectors). Our third (and last) definition of oriented matroids, which we are going to introduce below, does have this property. In fact, the socalled "complete" OMs, which will be defined in a minute, are such that one can recognize subspaces of $\mathbb{K}^E$ as oriented matroids at once.

5.9 Approximation Axioms

Let $L \leq \mathbb{K}^E$ be a subspace. In Example 5.20, we have shown that $\sigma(L)$ has the elimination property. There we considered $x, y \in L$ such that x and y were separated by an element $e \in E$ and we observed, that we can eliminate e between x and y by fixing some $f \in supp\, x \cup supp\, Y$ which does not separate x and y, thus obtaining the vector $z = \lambda x + \mu y$, λ and $\mu > 0$, with $z_e = 0$ and z_f having the sign of $x_f + y_f$. Actually, these arguments prove more than the simple fact that $\sigma(L)$ has the elimination property. Since the element $f \in E$ is arbitrary, they prove that z_f has the same sign as $x_f + y_f$ for every $f \in E$ which does not separate x and y. In terms of sign vectors, this can be stated as follows:

Proposition 5.45 *Let $L \leq \mathbb{K}^E$ be a subspace, and let $\mathcal{O} := \sigma(L)$. Then, given $X, Y \in \mathcal{O}$ and an element $e \in sep(X,Y)$, there exists a sign vector $Z \in \mathcal{O}$ such that $Z_e = 0$ and $Z_f = (X \circ Y)_f$ for every $f \in E \setminus sep(X,Y)$.*

□

Example 5.46 *Let $L := im\, A$ for some matrix $A \in \mathbb{K}^{E \times n}$ and $\mathcal{O} := \sigma(L)$. Furthermore, let $u, v \in \mathbb{K}^n$, and let $X := \sigma(Au)$, $Y := \sigma(Av)$. If $e \in sep(X,Y)$, this means that u and v lie on opposite sides of the hyperplane $H_e := \{w \in \mathbb{K}^n \mid A_{e.} \cdot w = 0\}$. Now let $w := H_e \cap [u,v]$. Then, obviously, $Z = \sigma(Aw)$ is as above. One might say, that we obtained w by* **approximating** *u and v on H_e. This terminology can be justified by observing that w minimizes the term $||w' - u|| + ||w' - v||$, $w' \in H_e$.*

Definition 5.47 *Let $X, Y \in 2^{\pm E}$ be two sign vectors, and let $e \in sep(X,Y)$. Then $Z \in 2^{\pm E}$ is said to be obtained by* **approximating** X **and** Y **on** e, *if $Z_e = 0$ and $Z_f = (X \circ Y)_f$ for every $f \in E \setminus sep(X,Y)$.*

A set $\mathcal{S} \subseteq 2^{\pm E}$ is said to have the **approximation property**, *if the following holds:*

(AP) *Given $X, Y \in \mathcal{S}$ and $e \in sep(X,Y)$, there exists $Z \in \mathcal{S}$ which is obtained by approximating X and Y on e.*

Note, that (AP) is essentially some kind of "strong" elimination property. If Z is obtained by approximating X and Y on e, this means that Z is obtained by "eliminating e between X and Y, fixing all elements f which do not separate X and Y". More precisely, the connection between the elimination and the approximation property is the following:

Lemma 5.48 *If $\mathcal{S} \subseteq 2^{\pm E}$ is symmetric and has the elimination property, then $span\, \mathcal{S}$ has the approximation property.*

Proof. Let $\mathcal{S}$ have the elimination property, and let $X, Y \in span\,\mathcal{S}$ be separated by some $e \in E$. Let us first assume that $X, Y \in \mathcal{S}$. Then, by assumption, for every $f \in supp\,X \cup supp\,Y$ which does not separate X and Y, there exists a $Z(f) \in \mathcal{S}$ which is obtained by eliminating e between X and Y, fixing f. The composition Z of all these sign vectors $Z(f)$ is an element of $span\,\mathcal{S}$ and it obviously approximates X and Y on e.

Next consider the general case $X, Y \in span\,\mathcal{S}$. Since $\mathcal{S}$ has the elimination property, Theorem 5.36 implies that X and Y can be written as conformal sums $X = X_1 \circ \ldots \circ X_k$ and $Y = Y_1 \circ \ldots \circ Y_l$ of vectors in $\mathcal{S}$. We proceed by induction on $k + l$. If $k = 1$ and $l = 1$, i.e. $X, Y \in \mathcal{S}$, the argument from above applies. Thus assume that, say, $k > 1$. Let $X' := X_1$ and $X'' := X_2 \circ \ldots \circ X_k$. By induction, there exist $Z', Z'' \in span\,\mathcal{S}$ which are obtained by approximating X' and Y, resp. X'' and Y on e. Then it is easy to see that $Z := Z' \circ Z''$ is an element of $span\,\mathcal{S}$ which is obtained by approximating X and Y on e.

□

We now give the definition of oriented matroids by means of approximation axioms. As we will see, the definition below characterizes the sets span$\,\mathcal{F}$, where $(\mathcal{F}, \mathcal{F}')$ is an OM.

Definition 5.49 *Let $\mathcal{O} \subseteq 2^{\pm E}$. Then $\mathcal{O}$ is called a* **complete oriented matroid** *("OM") if the following holds:*

a) *$0 \in \mathcal{O}$, and $\mathcal{O}$ is symmetric,*

b) *$\mathcal{O} = span\,\mathcal{O}$,*

c) *$\mathcal{O}$ has the approximation property.*

Recall that if $(\mathcal{F}, \mathcal{F}')$ is an OM, then $(span\,\mathcal{F},\ \mathcal{F}')$ is also an OM and $span\,\mathcal{F} = (\mathcal{F}')^{\perp}$ by Corollary 5.41, hence $span\,\mathcal{F}$ is maximal with this property. This justifies the notion "complete".

Theorem 5.50 *If $\mathcal{C}$ is an elementary OM, then $span\,\mathcal{C}$ is a complete OM.*

Proof. Follows from Lemma 5.48.

□

Thus in order to prove that all three definitions of oriented matroids are equivalent, we are left to verify that a complete oriented matroid $\mathcal{O}$ gives rise to an OM. Essentially, this means to show that the pair $(\mathcal{O}, \mathcal{O}^{\perp})$ has the FARKAS Property (FP). Thus we are left to give a proof of FARKAS' Lemma in the abstract setting of sign vectors.

5.10 Proof of FARKAS' Lemma in OMs

We are going to show that, given a complete oriented matroid $\mathcal{O}$, the pair $(\mathcal{O}, \mathcal{O}^\perp)$ is an OM. From Definition 5.8 of OMs, it is clear that we have to deal with minors of $(\mathcal{O}, \mathcal{O}^\perp)$. Recall that we defined a minor of a pair $(\mathcal{S}, \mathcal{S}')$ to be a pair $(\mathcal{S} \setminus I/J, \mathcal{S}' \setminus J/I)$. In other words, we *defined* the contraction to be the operation "dual" to deletion and vice versa. (Here "dual operation" means "corresponding operation on the orthogonal complement".) This has been justified by the fact that $(L \setminus I/J)^\perp = L^\perp \setminus J/I$ for every subspace $L \leq \mathbf{K}^E$. However, we did not *prove* the corresponding result for general sets of sign vectors!

Lemma 5.51 *Let $\mathcal{O} \subseteq 2^{\pm E}$ such that $\mathcal{O}$ is symmetric and $\mathcal{O}$ has the approximation property. Then for every $e \in E$ we have $(\mathcal{O} \setminus e)^\perp = \mathcal{O}^\perp/e$. In particular, this holds, if $\mathcal{O}$ is a complete OM.*

Proof. "$\supseteq$" follows immediately from the definition of minors. To prove the converse direction, let $\tilde{U} \in (\mathcal{O} \setminus e)^\perp$. We want to show that there exists $U \in \mathcal{O}^\perp$ such that $U_f = \tilde{U}_f$ for every $f \in E \setminus e$. Clearly, there are just three possibilities for U: Either $U_e = +$, $U_e = -$ or $U_e = 0$. We denote this by $U = \tilde{U} + e^+$, $U = \tilde{U} + e^-$ and $U = \tilde{U} + e^0$, resp. Assume that non of these is an element of $\mathcal{O}^\perp$. Hence there exist X, Y, Z in $\mathcal{O}$ such that $\tilde{U} + e^+ \not\perp X$, $\tilde{U} + e^- \not\perp Y$ and $\tilde{U} + e^0 \not\perp Z$. By reversing signs, if necessary, we may assume that the situation is as follows:

$$
\begin{array}{rcl}
 & & \qquad\qquad\qquad\qquad\quad e \\
\tilde{U} + e^+ & = & +\ \ \ldots + - \ldots - 0 \ldots 0\ + \\
\tilde{U} + e^- & = & +\ \ \ldots + - \ldots - 0 \ldots 0\ - \\
\tilde{U} + e^0 & = & +\ \ \ldots + - \ldots - 0 \ldots 0\ \ 0 \\
X & = & \oplus\ \ \ldots \oplus \ominus \ldots \ominus * \ldots *\ + \\
Y & = & \oplus\ \ \ldots \oplus \ominus \ldots \ominus * \ldots *\ - \\
Z & = & + \oplus \ldots \oplus \ominus \ldots \ominus * \ldots *\ \ *
\end{array}
$$

We may assume w.l.o.g. that $Z_e = 0$. For, if $Z_e \neq 0$, we may approximate Z and X or Z and Y (according to whether $Z_e = -$ or $Z_e = +$) on e in order to get a zero entry in coordinate e of Z. Thus we may in fact assume that $Z_e = 0$. But this immediately gives a contradiction, since then $\tilde{Z} = Z \setminus e \in \mathcal{O} \setminus e$ is not orthogonal to $\tilde{U} \in (\mathcal{O} \setminus e)^\perp$.

□

Surprisingly easy to prove is the following "dual" result:

Lemma 5.52 *Let $\mathcal{O} \subseteq 2^{\pm E}$ have the approximation property. Then for every $e \in E$ we have $(\mathcal{O}/e)^\perp = \mathcal{O}^\perp \setminus e$. In particular, this holds, if $\mathcal{O}$ is a complete OM.*

Proof. Again, "$\supseteq$" follows from the definition of minors. To prove "$\subseteq$", let $\tilde{Z} \in (\mathcal{O}/e)^\perp$ and define $Z \in 2^{\pm E}$ by $Z_e := 0$ and $Z_f := \tilde{Z}_f$ for every $f \in E \setminus e$. We have to show that $Z \in \mathcal{O}^\perp$. Thus, let $Y \in \mathcal{O}$, and let $\tilde{Y} = Y \setminus e \in \mathcal{O}/e$. Then $Y \perp Z$ follows immediately from $\tilde{Y} \perp \tilde{Z}$, which proves the claim.

□

Lemma 5.53 *Let $\mathcal{O}$ be a complete OM. Then every minor and every reorientation of $\mathcal{O}$ is a complete OM.*

Proof. This follows immediately from the definitions.

□

Thus, in order to show that $(\mathcal{O}, \mathcal{O}^\perp)$ is a an OM, it is sufficient to show that $(\mathcal{O}, \mathcal{O}^\perp)$ has the FARKAS Property. This can be done in exactly the same way as we proved FARKAS' Lemma in vector spaces:

Lemma 5.54 *Let $\mathcal{O} \subseteq 2^{\pm E}$ such that $\mathcal{O}$ is symmetric and $\mathcal{O}$ has the approximation property. Then for every $e \in E$ there exist $X \in \mathcal{O}$ and $Y \in \mathcal{O}^\perp$ such that $X \geq 0$ and $Y \geq 0$ and $(X \circ Y)_e = +$.*

Proof. The proof is by induction on $n = |E|$. The claim is trivially true for $n = 1$. Thus let $n = |E| \geq 2$ and assume the result holds for $n-1$. Let $e \in E$, and choose $f \neq e$, $f \in E$. We consider both $\mathcal{O} \setminus f$ and $\mathcal{O}/f$. These are complete OMs by Lemma 5.53. Applying our inductive assumption to $\mathcal{O} \setminus f$, we find (using Lemma 5.51) that

$$\begin{array}{rcll} & & \;e\;\,f & \\ \exists X & = & \oplus 0 \oplus \ldots \oplus & \in \mathcal{O} \\ \exists Y & = & \oplus * \oplus \cdots \oplus & \in \mathcal{O}^\perp \text{ with } (X \circ Y)_e = + \end{array}$$

Similarily, considering $\mathcal{O}/f$ (and using Lemma 5.52), we get

$$\begin{array}{rcll} \exists X' & = & \oplus * \oplus \ldots \oplus & \in \mathcal{O} \\ \exists Y' & = & \oplus 0 \oplus \cdots \oplus & \in \mathcal{O}^\perp \text{ with } (X' \circ Y')_e = + \end{array}$$

Since $X' \perp Y$, we see that X' and Y cannot both have a "$-$" entry in coordinate f. The claim follows.

□

Theorem 5.55 (FARKAS' Lemma) *Let $\mathcal{O}$ be a complete OM. Then $(\mathcal{O}, \mathcal{O}^\perp)$ has the FARKAS' Property. Hence, since complete OMs are closed under taking minors and reorientations, $(\mathcal{O}, \mathcal{O}^\perp)$ is an OM.*

□

5.11 Duality

The equivalence of the three definitions of oriented matroids essentially says, that if $\mathcal{F} \subseteq 2^{\pm E}$ has some "nice" properties (such as elimination or approximation property), then the orthogonal complement $\mathcal{F}^\perp$ is nice, too. This principle of "duality" or "orthogonality" is fundamental in the theory of oriented matroids. The duality theorems, listed below, are quite hard to prove from scratch, but they turn out to be immediate corollaries of our previous results.

Theorem 5.56 *If $\mathcal{O}$ is a complete OM, then so is $\mathcal{O}^\perp$. Furthermore, $\mathcal{O}^{\perp\perp} = \mathcal{O}$. ($\mathcal{O}^\perp$ is called the* **dual** *of $\mathcal{O}$.)*

Proof. If $\mathcal{O}$ is a complete OM, then $(\mathcal{O}, \mathcal{O}^\perp)$ is an OM by Theorem 5.55. Therefore, $elem\, \mathcal{O}^\perp$ is an elementary OM and $span(elem\, \mathcal{O}^\perp)$ is a complete OM. Since $(\mathcal{O}^\perp, \mathcal{O})$ is an OM, $(elem(\mathcal{O}^\perp), \mathcal{O})$ is an OM, too, by Corollary 5.38. Thus, Corollary 5.41 states that $\mathcal{O}^\perp$ equals $span(elem\, \mathcal{O}^\perp)$, and hence $\mathcal{O}^\perp$ is a complete OM. Furthermore, $\mathcal{O} = span\, \mathcal{O} = \mathcal{O}^{\perp\perp}$ by Corollary 5.41.

□

Theorem 5.57 *Let $\mathcal{C}$ be an elementary OM. Then $\mathcal{C}^\perp$ is a complete OM and $\mathcal{C}^* := elem\, \mathcal{C}^\perp$ is an elementary OM. Moreover, $\mathcal{C}^{**} = \mathcal{C}$.*

Proof. Let $\mathcal{C}$ be an elementary OM. Then $span\, \mathcal{C}$ and (by Theorem 5.56) $(span\, \mathcal{C})^\perp$ are complete OMs. Since $(span\, \mathcal{C})^\perp = \mathcal{C}^\perp$ (cf. Lemma 5.33), this proves the first part of the claim. Moreover, $\mathcal{C}^{**} = elem((elem(\mathcal{C}^\perp))^\perp) = elem(\mathcal{C}^{\perp\perp}) = elem(span\, \mathcal{C}) = elem\, \mathcal{C} = \mathcal{C}$.

□

In Chapter 7, where we shall see how to formulate "Linear programs" in OMs, duality will be studied more extensively. In most cases, we will deal with complete OMs only. So, for simplicity, let us make the convention that the term "OM" shall apply to all three kinds of oriented matroids. It will always be clear from the context or from notation ($\mathcal{C}$ or $\mathcal{O}$, resp $\mathcal{F}$) whether we consider elementary or complete OMs.

5.12 Further Reading

It is always difficult to say who has been the first. Among those who prepared the way to oriented matroids, one should mention

R.T. ROCKAFELLAR *The Elementary Vectors of a Subspace of $\mathbb{R}^n$*, in Combinatorial Mathematics and its Applications, pp. 104–127 in "Proc. of the Chapel Hill Conference", Chapel Hill (R.C. Bose and T.A. Dowling, eds.) University of North Carolina Press, Chapel Hill (1969).

G.J. MINTY *On the Abstract Foundations of the Theory of Directed Linear Graphs, Electrical Networks and Network Programming*, Journal of Mathematical Mech. **15** (1966), pp. 485–520.

Rockafellar, in the above mentioned article, seems to have foreseen much of the developement in oriented matroid theory. The paper of Minty presents the concept of digraphoids, an abstraction of duality in directed graphs, as we discussed it in Chapter 2. Oriented matroids, as we know them today, have been introduced independently by R. Bland, M. Las Vergnas, A. Dress, J. Folkman and J. Lawrence, cf.

R. G. BLAND *A Combinatorial Abstraction of Linear Programming*, Journal of Combinatorial Theory, Series bf B 23 (1977), pp. 33–57.

R. BLAND, M. LAS VERGNAS *Orientability of Matroids*, Journal of Combinatorial Theory, Series **B 24** (1978), pp. 94–123.

J. FOLKMAN, J. LAWRENCE *Oriented Matroids*, Journal of Combinatorial Theory, Series **B 25** (1978), pp. 199–236.

A. DRESS *Chirotopes and Oriented Matroids*, Bayreuther Mathematische Schriften **21** (1986), pp. 14–68.

The approach taken by Bland and Las Vergnas was to develop oriented matroids out of what is called "matroid theory", a theory that had been well established already at that time. Informally, matroid theory investigates systems of "sign vectors" having coordinates equal to "0" and "1" (representing a "$\neq 0$") only. Matroids arise as an abstraction of linear subspaces of $\mathbb{K}^n$ in the very same way as oriented matroids do if one does not only apply the "forgetting magnitudes map" $\sigma : \mathbb{K}^n \to \{0,+,-\}^n$, but rather a "forgetting magnitudes and signs map" $\tilde{\sigma} : \mathbb{K}^n \to \{0,1\}^n$. Thus, if one replaces the signs "+" and "−" by "1" in the definition of matroids, one arrives at a definition of matroids. For example, the elimination axiom yields the following characterization of matroids:

A matroid is a system $\zeta \subseteq \{0,1\}^E$ such that

(i) $0 \notin \zeta$

(ii) $\zeta = elem\, \zeta$

(iii) If $X, Y \in \zeta$ and $X_e = Y_e = 1$, $X_f = 1$, $Y_f = 0$, then exists $Z \in \zeta$ which is obtained by eliminating e betw X and Y, fixing f, i.e. $Z_e = 0$, $Z_f = 1$ and $supp\, Z$ $supp\, X \cup supp\, Y$.

The beginning of matroid theory dates back to the early work of Whitney, cf.

H. WHITNEY *The Abstract Properties of Linear Dependence*, American Journal of Mathematics **57** (1935), pp. 509–533.

Since 1960, the field has developed rapidly and has become one of the major research areas within combinatorial optimization.

From the above definition of matroids it is obvious that every oriented matroid has an underlying ("unoriented") matroid, that is obtained by "forgetting signs". Conversely, given an arbitrary matroid, one may try to "orient" it, i.e. replace every "1" entry in each vector by a "+" or "−" coordinate such that the resulting system of sign vectors is an oriented matroid. This is not always possible, i.e. there do exist "nonorientable" matroids. Nonetheless, it is clear that it is easier to develop oriented matroids out of matroid theory than starting from scratch (i.e. from linear subspaces of $\mathbb{K}^n$). We did not want to introduce matroid theory, however, since this is somewhat aside the other topics treated in this book, such as linear programming and polyhedral theory. Therefore, we preferred to choose the direct way to oriented matroids, which is abstract linear duality. This approach, which is the most convenient one for those not familiar with matroid theory, has been worked out by several authors, cf. [28], [70], [78].

Andreas Dress [60] developed in 1977 from a definition in terms of the Grassmann-Plücker-relations a new structure which he together with the organic chemist Andree Dreiding and his mathematical collaborator Hans Haegi used to investigate the chirality of organic molecules. Later on it turned out that these chirotopes are equivalent to oriented matroids. Moreover, Lawrence [121] showed that Gutierrez-Novoa's (cf. [98]) n-ordered sets are another variant of this approach. They derive oriented matroids by introducing an orientation on the bases of the underlying (unoriented) matroid. For those who are familiar with matroid theory, let us briefly outline the main idea of this approach.

Let E be a finite set, $d \leq |E|$ and let $det: E^d \to \{0, \pm 1\}$ be a nonvanishing alternating map. Then (E, det) is called a "base oriented" matroid, if for all $(f_1, \ldots, f_d) \in E^d$ and $(g_1, \ldots, g_d) \in E^d$ the following implication ("Grassmann-Plücker relation") holds:

$\forall j: \ det(g_j, f_2, \ldots, f_d) \cdot det(g_1, \ldots, g_{j-1}, f_1, g_{j+1}, \ldots, g_d) \geq 0$

implies

(ii) $\quad det(f_1, \ldots, f_d) \cdot det(g_1, \ldots, g_d) \geq 0\,.$

It is easy to see that the sets $\{b_1, \ldots, b_d\}$ for which $det(b_1, \ldots, b_d) \neq 0$ form a set $\mathcal{B}$ of bases of some matroid. Indeed, let $B = \{b_1, \ldots, b_d\} \in \mathcal{B}$ and $B' = \{b'_1, \ldots, b'_d\} \in \mathcal{B}$ and, say, $b_1 \in B \setminus B'$. Since

$$det(b_1, \ldots, b_d) \cdot det(b'_1, \ldots, b'_d) \neq 0\,,$$

the above condition implies that for some j we have

$$det(b'_j, b_2, \ldots, b_d) \neq 0\,.$$

Hence, $B \setminus b_1 \cup b'_j \in \mathcal{B}$. This shows that $\mathcal{B}$ is indeed the family of bases of a matroid M of rank d.

In the following we will briefly sketch how one can construct from a given base oriented matroid (E, det) the set of sign vectors of the corresponding elementary oriented matroid $\mathcal{C}$ and its dual $\mathcal{C}^*$. Details can be found in Lawrence [121].

Given a circuit C of M, let $B = \{b_1, \ldots, b_d\}$ be any basis such that C is a fundamental circuit with respect to B, i.e. $C \subseteq B \cup b_0$ for some element $b_0 \in E$. Define

$$\begin{aligned} X^+ &:= b_0 \cup \{b_i \mid det(b_1, \ldots, b_{i-1}, b_0, b_{i+1}, \ldots, b_d) &= -det(b_1, \ldots, b_d)\} \\ X^- &:= \{b_i \mid det(b_1, \ldots, b_{i-1}, b_0, b_{i+1}, \ldots, b_d) &= det(b_1, \ldots, b_d)\} \end{aligned}$$

Then $X^+ \dot\cup X^-$ is a partition of C. Using the Grassman-Plücker relations, one can show that this partition is independent from the particular choice of the basis B. The set $\mathcal{C}$ of sign vectors X defined this way is an elementary OM whose underlying matroid is M.

The dual $\mathcal{C}^*$ can be obtained from (E, det) as follows. Let C^* be a cocircuit of M and let $F = E \setminus C^*$ be the corresponding hyperplane. Let $\{f_2, \ldots, f_d\}$ be a basis of F. Then

$$\begin{aligned} Y^+ &:= \{e \mid det(e, f_2, \ldots, f_d) = +1\} \quad \text{and} \\ Y^- &:= \{e \mid det(e, f_2, \ldots, f_d) = -1\} \end{aligned}$$

form a partition of C^*. Again, one can show that this does not depend on the particular choice of $f_1, \ldots, f_{d-1}$. The set $\mathcal{C}^*$ of sign vectors Y obtained this way is the elementary OM dual to $\mathcal{C}$. Note that $\mathcal{C} \perp \mathcal{C}'$. In fact, let $X \in \mathcal{C}$ and $Y \in \mathcal{C}^*$ as above and assume that, say, $b_0 \in X^+ \cap Y^+$. Thus

$$det(b_0, f_2, \ldots, f_d) = +1\,.$$

Assume w.l.o.g. that, say, $det(b_1, \ldots, b_d) = +1$. Hence,

$$det(b_0, f_2, \ldots, f_d)\ det(b_1, \ldots, b_d) = +1 \ .$$

The Grassman-Plücker relations now imply that for some i, $1 \leq i \leq d$, we have

$$det(b_i, f_2, \ldots, f_d)\ det(b_1, \ldots, b_{i-1}, b_0, b_{i+1}, \ldots, b_d) = +1 \ .$$

Now it is obvious that $b_i \in X^+$ implies $b_i \in Y^-$ and, similarly, $b_i \in X^-$ implies $b_i \in Y^+$. Hence X and Y in fact orthogonal. In general it can be shown that two sets $\mathcal{C}$ and $\mathcal{C}^*$ of sign vectors form a dual pair of elementary OMs, provided the following two conditions are satisfied:

(i) $\mathcal{C} \perp \mathcal{C}^*$

(ii) $supp\,\mathcal{C} := \{supp\,X \mid X \in \mathcal{C}\}$ and $supp\,\mathcal{C}^* := \{supp\,Y \mid Y \in \mathcal{C}^*\}$

are the sets of circuits res. cocircuits of a matroid.

One should mention that from the viewpoint of matroid theory, oriented matroids are, though quite natural structures, not the only possible ones one can get out of matroids. For example, Bland has investigated a weaker notion of "orientability" of matroids, socalled "weakly oriented matroids", cf. [29].

The reader interested in matroid theory is referred to the following textbooks on that subject:

H. CRAPO, G.-C. ROTA *Combinatorial Geometries*, Preliminary Edition, MIT Press, Cambridge, Mass. (1970).

D.J.A. WELSH *Matroid Theory*, Academic Press (1976).

N. WHITE (ed.) *Theory of Matroids*, Cambridge University Press (1986).

As mentioned already, Folkman and Lawrence have independently discovered oriented matroids at about the same time as Bland and Las Vergnas had. In contrast to the approach of Bland and Las Vergnas which may be called the "combinatorial approach" to oriented matroids, Folkman and Lawrence's work is based on topological concepts. Their idea is to represent oriented matroids by means of "sphere systems". This trail of thought has been further persued by A. Mandel, cf.

A. MANDEL *Topology of Oriented Matroids*, Thesis, University of Waterloo, Canada (1981), supervised by J. Edmonds.

The characterization of oriented matroids in terms of sphere systems will be studied in Chapter 9.

Finally, let us again mention the contribution of A. Dress to the field. His approach to oriented matroids is what may be termed the "algebraic" one. He introduced the notion of socalled "matroids with coefficients" (cf. [58]), a concept that may be used to treat both ordinary matroids and oriented matroids in a unified framework. Essentially, the idea consists in replacing $\mathbb{K}^E$ by R^E, where R is a semiring with certain properties. This approach has been further investigated e.g. in [164], but will not be persued here.

Chapter 6

Linear Programming Duality

After this "tour de force" in abstract duality theory, the reader — if there still is any — will probably be glad to learn that we are now coming back to $\mathbb{K}^n$, in order to have a short break there and solve our optimization problems from Chapter 4. Our main object however will be to show that linear programming essentially is an oriented matroid problem.

6.1 The Dual Program

Consider a linear optimization problem

$$\max\{c^T x \mid Ax \leq b\} \tag{6.1}$$

defined over $\mathbb{K}$, i.e. A, b, c and x shall have entries in $\mathbb{K}$. To every $z \in \mathbb{K}$ we may associate the polyhedron $P_z \subseteq \mathbb{K}^n$, defined by the system of inequalities

$$\begin{aligned} Ax &\leq b \\ c^T x &\geq z \end{aligned}$$

as sketched in Figure 6.1 below. Thus $P_z = P\left(\binom{A}{-c^T}, \binom{b}{-z}\right)$.

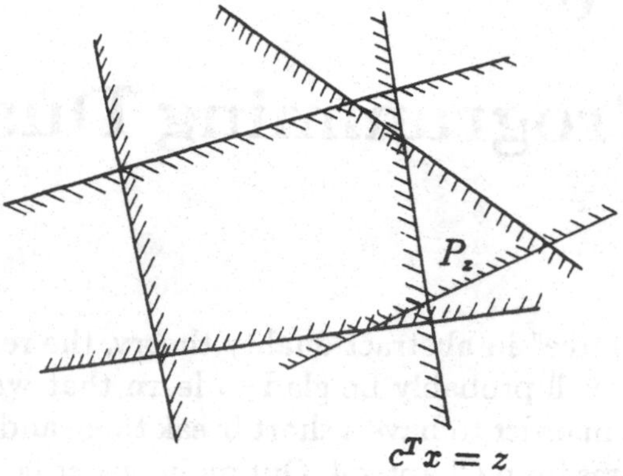

Figure 6.1

Let $z^* := \max\{c^T x \mid Ax \le b\}$ be the optimal value of (6.1). Thus $z^* = -\infty$ if (6.1) is infeasible, $z^* = +\infty$ if (6.1) is unbounded and $z^* \in \mathbb{R}$ otherwise (in fact, z^* will turn out to be in $\mathbb{K}$ in this case, as we will see below). Furthermore, z^* can obviously be determined as follows:

$$\begin{aligned} z^* &= \max\{z \in \mathbb{K} \mid P_z \neq \emptyset\} \qquad (6.2)\\ &= \min\{z \in \mathbb{K} \mid P_z = \emptyset\} \end{aligned}$$

(Here again, we note that "sup" and "inf" would be more appropriate, but we prefer "max" and "min" since they are commonly used in the context of linear programming.)

In Chapter 4 we learned, that to every P_z we may associate a polyhedron P'_z, such that

$$P_z \neq \emptyset \Leftrightarrow P'_z = \emptyset.$$

thus we may restate (6.2) as

$$\begin{aligned} z^* &= \min\{z \in \mathbb{K} \mid P'_z \neq \emptyset\} \qquad (6.3)\\ &= \max\{z \in \mathbb{K} \mid P'_z = \emptyset\}. \end{aligned}$$

Obviously, there is a striking similarity between (6.2) and (6.3). Thus a natural question to ask is: Does there exist a linear programming problem associated to (6.3) in the same way as (6.1) corresponds to (6.2)? The answer, as we will see in a minute, turns out to be positive: (6.3) *does* correspond to an LP, which is a *minimization* problem (as you might expect, since (6.2) and (6.3) arise from each other by replacing "max" by "min" and vice versa) and which has the same optimal value z^*. This program will be called the *dual* of (6.1). By the symmetry of the relation between (6.2) and (6.3) it is quite clear, that duality is a symmetric relation, i.e. the dual of the dual of (6.1) will be the orginal program (6.1) again. But this will be worked out in detail below.

First, let us seek for an explicit formulation of the LP associated to (6.3). Recall from Corollary 4.2, that

$$P'_z = P^{=}\left(\begin{pmatrix} A^T - c \\ b^T - z \end{pmatrix}, \begin{pmatrix} 0 \\ -1 \end{pmatrix}\right).$$

Thus $P'_z \neq \emptyset$, if and only if either

$$\exists \begin{pmatrix} u \\ 1 \end{pmatrix} \in \mathbf{K}_+^{m+1}, A^T u = c, b^T u < z \tag{6.4}$$

$$\text{or} \quad \exists \begin{pmatrix} v \\ 0 \end{pmatrix} \in \mathbf{K}_+^{m+1}, A^T v = 0, b^T v < 0\,. \tag{6.5}$$

Note, that by FARKAS' Lemma, (6.5) states that $P = P(A, b)$ is empty, i.e. (6.1) is infeasible. Let us rule out this case in what follows, by assuming that (6.1) is feasible. Then (6.3) can be stated more explicitely as follows:

$$\begin{aligned} z^* &= \min\{z \in \mathbf{K} \mid P'_z \neq \emptyset\} \qquad (6.6) \\ &= \min\{z \in \mathbf{K} \mid \exists\, u \in \mathbf{K}_+^m, A^T u = c, b^T u < z\} \\ &= \min\{b^T u \mid u \in \mathbf{K}_+^m, A^T u = c\}. \end{aligned}$$

This last reformulation is the one we were looking for.

Definition 6.1 *The* **dual** *of the LP*

$$\max\{c^T x \mid Ax \leq b\}$$

is defined to be the LP

$$\min\{b^T u \mid A^T u = c, u \geq 0\}.$$

If an LP is considered together with its dual, the former is also referred to as the **primal** *problem.*

As we have seen above, the primal problem $\max\{cx \mid Ax \leq b\}$ and its dual $\min\{b^T u \mid A^T u = c, u \geq 0\}$, define the same optimal value z^*. — Well, not quite .... Recall that we have shown this under the additional assumption, that one of them (in fact, the primal one) is feasible. However, it may happen that both programs are infeasible, in which case their optimal values are $-\infty$ and $+\infty$, resp..

Example 6.2 *Let* $A = \begin{pmatrix} 1 & -1 \\ -1 & 1 \end{pmatrix}$, $b = \begin{pmatrix} -1 \\ -1 \end{pmatrix}$ *and* $c = \begin{pmatrix} 1 \\ 0 \end{pmatrix}$.

Then both $\max\{cx \mid Ax \le b\}$ *and* $\min\{b^T u \mid A^T u = c, u \le 0\}$ *are easily seen to be infeasible.*

As we remarked earlier, the symmetry between the relations (6.2) and (6.3) suggests, that duality between LPs is a symmetric relation, too. We can not prove this, however, before having defined the "dual" of a minimization problem. Thus consider the LP $\min\{cx \mid Ax \le b\}$. What should be a natural definition for its dual? Of course, we could go along the same way as above, working through all the P_z- and P'_z-reformulations. The reader is invited to do so — or just to believe that the following simple reasoning arrives at the same end:

Note that

$$\begin{aligned} z^* &:= \min\{c^T x \mid Ax \le b\} \\ &= -\max\{-c^T x \mid Ax \le b\} \\ &= -\min\{b^T u \mid A^T u = -c, u \ge 0\} \\ &= \max\{b^T u \mid A^T u = c, u \le 0\}. \end{aligned}$$

Definition 6.3 *The* **dual** *of the LP*

$$\min\{c^T x \mid Ax \le b\}$$

is defined to be the LP

$$\max\{b^T u \mid A^T u = c, u \le 0\}.$$

Now we are ready to prove explicitely the symmetry of the duality relation, i.e. to show that the dual of the dual is the original (primal) LP again: Consider $\min\{b^T u \mid A^T u = c, u \ge 0\}$, i.e. the dual of (6.1) and rewrite it as

$$\min\left\{ b^T u \;\middle|\; \begin{pmatrix} A^T \\ -A^T \\ -I \end{pmatrix} u \le \begin{pmatrix} c \\ -c \\ 0 \end{pmatrix} \right\},$$

where I denotes the unit matrix. The dual of this is

$$\max\left\{ (c^T, -c^T, 0) \begin{pmatrix} x^+ \\ x^- \\ x^0 \end{pmatrix} \;\middle|\; (A, -A, -I) \begin{pmatrix} x^+ \\ x^- \\ x^0 \end{pmatrix} = b \text{ and } x^+, x^-, x^0 \le 0 \right\}.$$

Substituting $x := x^+ - x^-$, this is easily seen to be eqivalent to

$$\max\{c^T x \mid Ax \le b\},$$

which is our original Problem (6.1).

We may summarize our results as follows

Theorem 6.4 *To every linear optimization problem (defined over $\mathbb{K}$) we may associate a dual problem as done in Definitions 6.1 and 6.3. The dual of the dual of an LP is the original LP again. Moreover, each of a dual pair of programs defines the same optimal value $z^* \in \mathbb{R} \cup \{+\infty, -\infty\}$, unless both problems are infeasible. Thus, in particular, if one of them is unbounded, the other is infeasible and if $\bar{x}$ and $\bar{u}$ are feasible solutions of the primal (say, a maximization) problem and its dual, resp., then $c^T\bar{x} \leq z^* \leq b^T\bar{u}$. Hence both are optimal, if $c^T\bar{x} = b^T\bar{u}$.*

After all, the reader may have got the impression that duality of LPs is something much more complicated than duality between linear vector spaces or oriented matroids. However, it is not. As we will see in the next section, things become much easier in the abstract setting. In particular, the definition of dual matroid problems will be much more attractive than Definitions 6.1 and 6.3 in that it reflects more of the symmetry principle of duality — as was still apparent in (6.2) and (6.3). Nevertheless, we decided to introduce dual programs in the way we did, for two reasons. First, you will find Definitions 6.1 and 6.3 in every introductory book on linear programming, and we did not want you to miss them here. Secondly, whenever you are to solve Linear Programming problems, you will have to deal with matrices rather than oriented matroids. By the way: Note that Theorem 6.4 again provides a simple stopping criterion for any LP-algorithm: Recall that in Section 3.2 our problem has been to check whether or not a given feasible solution is optimal. Suppose that x is a feasible solution of $\max\{c^Tx \mid Ax \leq b\}$ with objective function value $z^* = c^Tx$. Then Theorem 6.4 tells us, that for proving optimality of x, it suffices to exhibit a feasible solution u of the dual problem having the same objective function value $z^* = b^Tu$, showing that x and u are in fact optimal solutions of the primal and dual program, resp.. Furthermore, in this case the optimal value $z^* = c^Tx = b^Tu$ is clearly an element of $\mathbb{K}$. We will prove in the next section, that such a dual pair of optimal solutions always exists, provided both problems are feasible.

Let us finish this section by considering the dual of the Diet problem (cf. Example 3.2).

Example 6.5 (The Pill-Maker's Problem) *Writing the Diet Problem in the form*

$$\min c^Tx$$
$$\binom{-A}{-I}x \leq \binom{-r}{0}$$

we see that the dual is

$$\max\,(-r^T,0)\binom{u}{v}$$
$$(u^T,v^T)\binom{-A}{-I}=c$$
$$u,v\le 0$$

or, equivalently,

$$\max\, r^Tu$$
$$u^TA\le c$$
$$u\ge 0.$$

This has the following interpretation: A pill-maker wants to market pills, each containing one of the m nutrients, at a price of u_i per unit of nutrient i. He wishes to be competetive with the price of real food, while at the same time maximizing the cost of an adequate diet. Indeed, the constraint $u^TA_{.j}\le c_j$ expresses the fact, that the cost in pill form of all the nutrients contained in the j-th food does not exceed the cost of the j-th food itself. The objective r^Tu is simply the cost of an adequate diet (i.e. the money he gets from the homemaker, when selling his pills to him).

Note, that in this problem A, r and c are nonnegative. Thus both the Diet problem and its dual will be feasible (since $x=0$ and $u=0$ are feasible solutions). Therefore, Theorem 6.5 tells us that the homemaker's primal and the pill-maker's dual both determine the same optimal value $z^*\in\mathbb{R}$. In fact, they are really two ways of stating the same problem.

6.2 The Combinatorial Problem

In this section we want to present a more "combinatorial" version of the linear optimization problem

$$\max\, c^Tx \qquad\qquad (6.1)$$
$$Ax\le b$$

We will transform this problem in several steps such that, finally, the feasible solutions will become elements of an oriented matroid. Hence, in particular, the number of feasible solutions will become finite!

To begin with, let us reformulate the above program as follows:

$$\max\,(c^T,-c^T)\binom{x^+}{x^-}$$
$$(A,-A)\binom{x^+}{x^-}\le b$$
$$x^+,x^-\ge 0.$$

This is easily seen to be an equivalent formulation, by simply substituting $x := x^+ - x^-$. Thus, we may restrict ourselves to considering problems of the form

$$\begin{aligned} \max c^T x & \\ Ax \leq b & \\ x \geq 0. & \end{aligned} \tag{6.7}$$

The next step in our transformation process will be to introduce so-called **slack variables**, in order to obtain an equality $Ax = b$ instead of an inequality:

$$\begin{aligned} \max\, (c^T, 0^T)\binom{x}{s} & \\ (A, I)\binom{x}{s} = b & \\ x, s \geq 0. & \end{aligned}$$

This is again easily seen to be an equivalent formulation of (6.7), and hence we may restrict our attention to programs of the from

$$\begin{aligned} \max c^T x & \\ Ax = b & \\ x \geq 0. & \end{aligned} \tag{6.8}$$

(This is usually called the **Standard Form** of an LP.)

Again, (6.8) may be restated as

$$\begin{aligned} \max\, (0, c^T)x & \\ (-b, A)x = 0 & \\ x_0 = 1,\ x_1, \ldots, x_n \geq 0. & \end{aligned}$$

The advantage of this representation is, that our feasible solutions $x \in \mathbf{K}^{n+1}$ now correspond to elements of a vector space (and hence also to sign vectors of an oriented matroid). However, it is not clear, how to interprete the objective function $c^T x$ in this context. Therefore, our final step consists in "hiding the vector c among the contraints":

$$\max x_{n+1}$$

$$\begin{aligned} \begin{pmatrix} 0 & c^T & -1 \\ -b & A & 0 \end{pmatrix} x & = 0 \\ x_0 = 1,\ x_1, \ldots, x_n & \geq 0 \end{aligned}$$

If we define the $(m+1)\times(n+2)$ matrix $B := \begin{pmatrix} 0 & c^T & -1 \\ -b & A & 0 \end{pmatrix}$, then our problem finally becomes

$$\begin{aligned} &\max x_{n+1} \\ &Bx = 0 \\ &x_0 = 1,\ x_1,\dots,x_n \geq 0. \end{aligned} \tag{6.9}$$

Let us derive a similar representation of the dual problem. In order to do so, we write (6.9) in the form $\max\{c^T x \mid Ax \leq b\}$. This looks as follows:

$$\max (0,0,\dots,0,1)x$$

$$\begin{pmatrix} 1 & 0 & \cdots & \cdots & \cdots & \cdots & 0 & 0 \\ -1 & 0 & \cdots & \cdots & \cdots & \cdots & 0 & 0 \\ 0 & -1 & 0 & & & & 0 & 0 \\ \vdots & \vdots & -1 & \ddots & & & \vdots & \vdots \\ \vdots & \vdots & & \ddots & \ddots & & \vdots & \vdots \\ \vdots & \vdots & & & \ddots & \ddots & \vdots & \vdots \\ \vdots & \vdots & & & & \ddots & \ddots & \vdots \\ 0 & 0 & \cdots & \cdots & \cdots & 0 & -1 & 0 \\ & & & B & & & & \\ & & & -B & & & & \end{pmatrix} x \leq \begin{pmatrix} 1 \\ -1 \\ 0 \\ \vdots \\ \vdots \\ \vdots \\ \vdots \\ \vdots \\ \cdot \\ 0 \end{pmatrix}.$$

The dual of this is

$$\min r_1 - r_2$$

$$(r_1, r_2, s^{0^T}, s^{+^T}, s^{-^T}) \begin{pmatrix} 1 & 0 & \cdots & \cdots & \cdots & \cdots & 0 \\ -1 & 0 & \cdots & \cdots & \cdots & \cdots & 0 \\ 0 & -1 & 0 & \cdots & \cdots & \cdots & 0 \\ \vdots & \vdots & \ddots & \ddots & & & \vdots \\ \vdots & \vdots & & \ddots & \ddots & & \vdots \\ \vdots & \vdots & & & \ddots & \ddots & \vdots \\ 0 & 0 & \cdots & \cdots & \cdots & -1 & 0 \\ & & & & B & & \\ & & & & -B & & \end{pmatrix} = (0,\dots,0,1)$$

$$(r_1, r_2) \in \mathbf{K}_+^2, \quad s^0 \in \mathbf{K}_+^n, \ s^+ \in \mathbf{K}_+^{m+1}, \ s^- \in \mathbf{K}_+^{m+1}.$$

Substituting $u_0 = r_1 - r_2$, $u := s^+ - s^-$, we get

$$\min u_0$$

$$(u_0, u)^T \begin{pmatrix} 1 & 0 & \dots & 0 \\ & & B & \end{pmatrix} \begin{cases} = 0 & \leftarrow \text{first component} \\ \geq 0 & \leftarrow \text{intermediate components} \\ = 1 & \leftarrow \text{last component} \end{cases}$$

$$u \in \mathbf{K}^{m+1}.$$

This can also be written as

$$\min -y_0$$
$$u^T B = (y_0, \dots, y_{n+1})$$
$$y_1, \dots, y_n \geq 0, \ y_{n+1} = 1,$$

i.e.

$$\min y_0 \tag{6.10}$$
$$u^T B = y^T$$
$$y_1, \dots, y_n \leq 0, \ y_{n+1} = -1.$$

If we let $L := ker\, B$ and hence $L^\perp = im\, B^T$, then (6.9) and (6.10) become, resp.:

$$\max x_{n+1} \tag{6.11}$$
$$x \in L$$
$$x_0 = 1, \ x_1, \dots, x_n \geq 0$$

and

$$\min y_0 \tag{6.12}$$
$$y \in L^\perp$$
$$y_1, \dots, y_n \leq 0 \quad y_{n+1} = -1.$$

This is perhaps the most illuminating way to represent a pair of dual programs. (Note that these are in fact Linear Programs, since the conditions "$x \in L$" and "$y \in L^\perp$" may be expressed in terms of linear equalities.) In particular, one deduces immediatly the following socalled "weak complementary slackness theorem":

Theorem 6.6 *Let x and y be feasible solutions of (6.11) and (6.12), resp. Then x and y are optimal, if $supp\, x \cap supp\, y \subseteq \{0, n+1\}$.*

Proof. Let $x \in L$ and $y \in L^{\perp}$ be such that $x_0 = 1$, $y_{n+1} = -1$ and $supp\, x \cap supp\, y \subseteq \{0, n+1\}$. Then $0 = x^T y = x_0 y_0 + x_{n+1} y_{n+1} = y_0 - x_{n+1}$. Thus $x_{n+1} = y_0$, i.e. the two feasible solutions x and y have the same objective function value and hence their optimality follows from Theorem 6.4.

□

It is obvious from (6.11) and (6.12), how to define Linear Programming problems in oriented matroids. Every feasible solution x of (6.11) corresponds to a sign vector $X \in \mathcal{O} = \sigma(L)$. In fact this correspondence is even one-one for elementary vectors of L (note that scalar multiples are ruled out by the condition $x_0 = 1$). Thus, we may, more or less mechanically, and without thinking about its meaning, write down the following "oriented matroid program":

$$\begin{gathered} \max X_{n+1} \\ X \in \mathcal{O} \\ X_0 = +,\ X_1, \ldots, X_n \geq 0 \end{gathered}$$

However, it is far from being obvious, what should be meant by "$\max X_{n+1}$". Note that X_{n+1} is nothing but a sign, i.e. $+, -$ or 0. Hence, given two "feasible" solutions $X, X' \in \mathcal{O}$, both having "objective function value" equal to, say, "+", how can we say whether X' is "better" than X or not? Indeed, this is a very difficult question (causing a lot of problems, when oriented matroid optimization is considered from an algorithmic point of view). Fortunately, we will have no need to answer it. All we want to know is: Given a feasible solution $X \in \mathcal{O}$, is it "optimal" or not? This is much easier to answer. Let us look at "ordinary" LPs first:

Suppose we are given a feasible solution $x = (x_0, \ldots, x_{n+1})$ of (6.11). If $z \in L$ such that $z_0 = 0$, $z_{n+1} = 1$ and $z_i \geq 0$ whenever $x_i = 0 \quad (i = 1, \ldots, n)$, then we say that z is **improving** (with respect to x). Clearly, if $z \in L$ is improving with respect to x, then, for sufficiently small $\epsilon \in \mathbb{K}$, $\epsilon > 0$, $x' = x + \epsilon z$ will (still) be a feasible solution of (6.11), having objective function value equal to $x'_{n+1} = x_{n+1} + \epsilon$. Thus x can not have been optimal in this case. Conversely, assume that x is not optimal. Then, by definition, there exists a feasible solution x' of (6.11) such that $x'_{n+1} > x_{n+1}$. Let $z' := x' - x$. Then $z \in L$, $z'_0 = 0$, $z'_{n+1} > 0$ and $z'_i \geq 0$ whenever $x_i = 0$. Thus, up to scaling by a factor $1/z'_{n+1}$, z' is improving with respect to x. We have shown that x is optimal if and only if there is no $z \in L$ which is improving with respect to x. This notion readily carries over to general oriented matroid programs: A "feasible" solution $X \in \mathcal{O}$ will be called "optimal" if there is no $Z \in \mathcal{O}$ which is "improving" with respect to X.

Similarily, the notion of unboundedness can be defined for general oriented matroid programs. Suppose there exists a $z \in L$ such that $z_0 = 0$, $z_{n+1} = 1$ and $z_i \geq 0$ for all $i = 1, \ldots, n$ (hence z is improving with respect to every feasible x). Then, if (6.11) is feasible, say, $x \in L$ is a feasible solution, we see that $x + \epsilon z$ is feasible for every $\epsilon > 0$. Thus, in particular, (6.11) is unbounded.

We are now facing a long and boring definition.

Definition 6.7 *Let $\mathcal{O}$ be an OM on $E = \{0, \ldots, n+1\}$. Then $X \in \mathcal{O}$ is called a* **(primal) feasible solution,** *if $X_0 = +$ and $X_1, \ldots, X_n \geq 0$. $Y \in \mathcal{O}^\perp$ is called a* **(dual) feasible solution,** *if $Y_1, \ldots, Y_n \leq 0$ and $Y_{n+1} = -$. Furthermore, $\mathcal{O}(\mathcal{O}^\perp)$ is called* **feasible,** *if there exists a feasible solution $X \in \mathcal{O}$ $(Y \in \mathcal{O}^\perp)$, otherwise it is called* **infeasible.** *Given a feasible solution $X \in \mathcal{O}$, we say that $Z \in \mathcal{O}$ is* **improving** *(with respect to X), if $Z_0 = 0$, $Z_{n+1} = +$ and $Z_i \geq 0$, whenever $X_i = 0$ $(i = 1, \ldots, n)$. Similarly, given a feasible $Y \in \mathcal{O}^\perp$, we say that $Z \in \mathcal{O}^\perp$ is* **improving** *(with respect to Y), if $Z_0 = -$, $Z_{n+1} = 0$ and $Z_i \leq 0$ whenever $Y_i = 0$ $(i = 1, \ldots, n)$. A feasible solution $X \in \mathcal{O}$ $(Y \in \mathcal{O}^\perp)$ is* **optimal,** *if there is no $Z \in \mathcal{O}$ $(Z \in \mathcal{O}^\perp)$ that is improving with respect to X (Y). If X and Y are feasible solutions of $\mathcal{O}$ and $\mathcal{O}^\perp$, resp., then X and Y are called* **complementary,** *provided $\mathrm{supp}\, X \cap \mathrm{supp}\, Y \subseteq \{0, n+1\}$. Finally, $\mathcal{O}$ is called* **unbounded** *provided it is feasible and there exists a $Z \in \mathcal{O}$ such that $Z_0 = 0$, $Z_{n+1} = +$ and $Z_i \geq 0$ for all $i = 1, \ldots, n$. $\mathcal{O}^\perp$ is called* **unbounded,** *if it is feasible and there exists $Z \in \mathcal{O}^\perp$ such that $Z_0 = -$, $Z_{n+1} = 0$ and $Z_i \leq 0$ for all $i = 1, \ldots, n$. $\mathcal{O}$ $(\mathcal{O}^\perp)$ is called* **bounded,** *provided it is feasible and not unbounded.*

A dual pair of oriented matroid LPs *will be denoted by*

$$\begin{array}{lcl} \max X_{n+1} & & \min Y_0 \\ X \in \mathcal{O} & \text{and} & Y \in \mathcal{O}^\perp \\ X_0 = +, X_1, \ldots, X_n \geq 0 & & Y_1, \ldots, Y_n \leq 0, Y_{n+1} = - \end{array}$$

or, even more schematically,

$$X = + \oplus \ldots \oplus \uparrow \in \mathcal{O} \qquad \text{and} \qquad Y = \downarrow \ominus \ldots \ominus - \in \mathcal{O}^\perp.$$

where "$\oplus$" and "$\ominus$" indicate nonnegative, resp. nonpositive components and the arrows "$\uparrow$" resp. "$\downarrow$" indicate the component which is to be maximized resp. minimized.

Proposition 6.8 *Let $E = \{0, \ldots, n+1\}$, $L \leq \mathbb{K}^E$ and consider the dual pair of LPs given in (6.11) and (6.12). Let $\mathcal{O} = \sigma(L)$ and hence $\mathcal{O}^\perp = \sigma(L^\perp)$. Then every feasible solution x of (6.11) corresponds to* *a feasible solution $X = \sigma(x)$ of $\mathcal{O}$ and every feasible solution y of (6.12) corresponds to a feasible solution $Y = \sigma(y)$ of $\mathcal{O}^\perp$. Moreover, this correspondence is $1-1$ for elementary vectors $x \in L$ resp. $y \in L^\perp$. $x \in L$ ($y \in L^\perp$) is an optimal solution, if and only if the corresponding sign vector is optimal. Furthermore, (6.11) resp. (6.12) is feasible, infeasible, bounded or unbounded if and only if $\mathcal{O}$ resp. $\mathcal{O}^\perp$ is.*

Proof. Most of this is either trivial or has already been shown by our arguments preceding Definition 6.7. What is left, is to show that $\mathcal{O}$ is unbounded if and only if (6.11) is. We mentioned above, that one direction is easy. At the risk of repeating ourselves: Suppose that $\mathcal{O}$ is unbounded. Then by definition, $\mathcal{O}$ is feasible, say $X \in \mathcal{O}$ is a feasible solution, and there exists $Z \in \mathcal{O}$ such that $Z_0 = 0$, $Z_{n+1} = +$ and $Z_i \geq 0$ for all $i = 1, \ldots, n$. Let x and z be corresponding elements in L. Then $x + \varepsilon z$ is feasible for all $\epsilon \geq 0$, showing, that (6.11) is unbounded. The converse is, in fact, nontrivial: Suppose that (6.11) is unbounded. Then, by Theorem 6.4, (6.12) is infeasible. Thus, $\mathcal{O}^\perp$ is infeasible, i.e.

$$\nexists Y = * \ominus \ldots \ominus -, \quad Y \in \mathcal{O}^\perp$$

Applying MINTY's Lemma, this yields

$$\exists Z = 0 \oplus \ldots \oplus +, \quad Z \in \mathcal{O},$$

as required.

□

Let us summarize, what we have done so far. We started out with the general "ordinary" LP (6.1) and transformed it, together with its dual until we arrived at (6.11) and (6.12). All of these single transformation steps were such that an obvious correspondence between the optimal (non optimal) solutions of the various reformulations was maintained. Thus, in particular, (6.1) is feasible, unbounded, etc., has or does not have optimal solutions if and only if (6.11) does. Proposition 6.8 says, that a similar relationship holds between (6.11),(6.12) and the corresponding oriented matroid LPs. This time, however, we have not just reformulated the problem, but we have also reduced the set of feasible solutions to a finite set! Therefore, the reader may wonder about whether our original optimization problem has not become trivial in a certain sense. For example, suppose we are given a feasible solution x of (6.11) and we want to decide whether it is optimal or not. All we have to do is, to look at $\mathcal{O} = \sigma(L)$ and check whether among the (finitely many!) elements of $\mathcal{O}$ there is one which is improving with respect to $X = \sigma(x)$. Of course, this would be an easy thing, once we are given $\mathcal{O}$ explicitely, i.e. a list of all its sign vectors. However, in general, we are not (in fact, deciding whether a given sign vector $Z \in 2^{\pm E}$ is an element of $\mathcal{O} = \sigma(L)$, turns out to be as difficult as solving "ordinary" LPs.[1]) Thus, oriented matroids do not seem to be useful for solving LPs. They are, however, quite useful in that they provide us a means for proving LP-duality theorems in a very clear and systematic way, revealing precisely those ("combinatorial") properties of $\mathbb{K}^n$ that make the whole theory work.

Let us start our investigation of oriented matroid LPs by introducing some further notation. Sometimes it will be convenient to consider such Linear Programs for minors of $\mathcal{O}$ as well. If, e.g. $I = \{1, \dots, k\}$, $J = \{k+1, \dots, l\}$, then the LP associated to $\mathcal{O} \setminus I / J$ and its dual will be denoted by

$$X = +\overbrace{0\dots0}^{I}\overbrace{*\dots*}^{J}\oplus\dots\oplus\uparrow \in \mathcal{O}$$

resp.

$$Y = \downarrow *\dots*\ 0\dots0\ominus\dots\ominus - \in \mathcal{O}^{\perp}.$$

Furthermore, one may also consider oriented matroid LPs associated to reorientations of $\mathcal{O}$ or a minor of $\mathcal{O}$. For example, if I, J, K are disjoint subsets of $\{1, \dots, n\}$ then the LP corresponding to ${}_K\mathcal{O} \setminus I/J$ will be denoted by

$$X = +\overbrace{0\dots0}^{I}\overbrace{*\dots*}^{J}\overbrace{\ominus\dots\ominus}^{K}\oplus\dots\oplus\uparrow \in \mathcal{O}$$

and its dual is

$$Y = \downarrow *\dots*\ 0\dots0\oplus\dots\oplus\ominus\dots\ominus - \in \mathcal{O}^{\perp}.$$

Of course, we may also reorient coordinates 0 and $n+1$. For example, if in the last problem above coordinate $n+1$ is reoriented, then this turns the maximization problem into a minimization problem. Therefore, the LP that is obtained from the above by reorienting coordinate $n+1$ will be denoted by

[1]Indeed, $Z \in 2^{\pm E}$ is an element of $\mathcal{O} = \sigma(L)$, if and only if each of the following LPs

$\max_{x \in L} x_i$		$(i \in Z^+)$		$\min_{x \in L} x_i$		$(i \in Z^-)$
$x_j \geq 0$	$\forall j \in Z^+$		and	$x_j \geq 0$	$\forall j \in Z^+$	
$x_j \leq 0$	$\forall j \in Z^-$			$x_j \leq 0$	$\forall j \in Z^-$	
$x_j = 0$	otherwise			$x_j = 0$	otherwise,	

has an optimal solution with objective function value $x_i > 0$, if $i \in Z^+$ and $x_i < 0$, if $i \in Z^-$. In this case, if z denotes the sum over all these optimal solutions, $Z = \sigma(z) \in \mathcal{O}$.

$$X = +\overbrace{0\dots0}^{I}\overbrace{*\dots*}^{J}\overbrace{\ominus\dots\ominus}^{K}\oplus\dots\oplus\downarrow\in\mathcal{O}$$

and its dual will then be

$$Y = \downarrow *\dots*\ 0\dots0\ \oplus\dots\oplus\ominus\dots\ominus + \in \mathcal{O}^{\perp}.$$

Finally, we will sometimes allow other coordinates to play the role of coordinates 0 and $n+1$. This probably needs no further explanation. As an example consider the following LP

$$X = 0\dots0 *\dots* \uparrow \ominus\dots\ominus + \oplus\dots\oplus \in \mathcal{O}$$

and its dual

$$Y = *\dots*0\dots0 - \oplus\dots\oplus \downarrow \ominus\dots\ominus \in \mathcal{O}^{\perp}.$$

The following lemma provides some simple facts about oriented matroid programs:

Lemma 6.9 *Let $\mathcal{O}$, $\mathcal{O}^{\perp}$ be a dual pair of OMs on $E = \{0,\dots,n+1\}$, and consider the two dual oriented matroid LPs*

$$X = +\oplus\dots\oplus\uparrow\in\mathcal{O} \quad \text{and} \quad Y = \downarrow\ominus\dots\ominus - \in\mathcal{O}^{\perp}.$$

Then the following holds:

(i) *A feasible solution $X \in \mathcal{O}$ is optimal if and only if there exists a feasible solution $Y \in \mathcal{O}^{\perp}$ such that X and Y are complementary. Dually, a feasible solution $Y \in \mathcal{O}^{\perp}$ is optimal if and only if there exists a feasible complementary $X \in \mathcal{O}$.*

(ii) *If $X \in \mathcal{O}$ is feasible, and $\bar{X} \in \mathcal{O}$ is optimal, then $\bar{X}_{n+1} \geq X_{n+1}$. (Here "$\geq$" denotes the obvious ordering on $\{+,-,0\}$, i.e. $- \leq 0 \leq +$.) In particular, if X and X' are optimal solutions, then $X_{n+1} = X'_{n+1}$. Again, the dual statement, i.e. the corresponding statement about feasible and optimal solutions $Y \in \mathcal{O}^{\perp}$ is also true.*

(iii) *If $X \in \mathcal{O}$ and $Y \in \mathcal{O}^{\perp}$ are primal/dual optimal solutions, then $X_{n+1} = Y_0$.*

(iv) *If $\mathcal{O}$ is feasible, then $\mathcal{O}$ is unbounded if and only if $\mathcal{O}^{\perp}$ is infeasible. Dually, if $\mathcal{O}^{\perp}$ is feasible, then $\mathcal{O}^{\perp}$ is unbounded if and only if $\mathcal{O}$ is infeasible.*

Proof.

ad (i): Let $X \in \mathcal{O}$ be feasible. Then, by definition, X is optimal if and only if there is no $Z \in \mathcal{O}$ which is improving with respect to X. Perhaps the situation is most easily understood if we illustrate it as follows: Suppose that the elements of E have been reordered such that $X_1 = \ldots = X_k = +$ and $X_{k+1} = \ldots = X_n = 0$ $(0 \leq k \leq n)$, i.e.

$$\begin{aligned} \exists X &= ++\ldots+0\ldots0* \in \mathcal{O} \\ \nexists Z &= 0*\ldots*\oplus\ldots\oplus+ \in \mathcal{O}. \end{aligned}$$

By MINTY's Lemma (and since $\mathcal{O} = -\mathcal{O}$), the latter is equivalent to

$$\exists Y = *0\ldots0\ominus\ldots\ominus- \in \mathcal{O}^\perp$$

which is a feasible solution of $\mathcal{O}^\perp$ and complementary to X. This proves the first claim of (i). The dual statement can be derived in the same way.

ad (ii): Let $X \in \mathcal{O}$ be feasible, and let $\bar{X} \in \mathcal{O}$ be optimal. By (i), there exists a dual feasible $\bar{Y} \in \mathcal{O}^\perp$ which is complementary to $\bar{X}$. Suppose first that $\bar{X}_{n+1} = 0$. Then (by appropriately ordering the elements $1, \ldots, n$), we get:

$$\begin{aligned} X &= +\oplus \quad \ldots\ldots \quad \oplus * \in \mathcal{O}, \\ \bar{X} &= +\oplus\ldots\oplus\ 0\ \ldots\ 0\ 0 \in \mathcal{O}, \\ \bar{Y} &= *\ 0\ \ldots\ 0\ \ominus\ldots\ominus - \in \mathcal{O}^\perp. \end{aligned}$$

Now $\bar{X} \perp \bar{Y}$ implies $\bar{Y}_0 = 0$. Hence $X \perp \bar{Y}$ implies $X_{n+1} = \ominus$, i.e. $X_{n+1} \leq \bar{X}_{n+1}$. Next assume that $\bar{X}_{n+1} = -$. Then

$$\begin{aligned} X &= +\oplus \quad \ldots\ldots \quad \oplus * \in \mathcal{O}, \\ \bar{X} &= +\oplus\ldots\oplus\ 0\ \ldots\ 0\ - \in \mathcal{O}, \\ \bar{Y} &= *\ 0\ \ldots\ 0\ \ominus\ldots\ominus - \in \mathcal{O}^\perp. \end{aligned}$$

Now $\bar{X} \perp \bar{Y}$ implies $\bar{Y}_0 = -$, while $X \perp \bar{Y}$ implies $X_{n+1} = -$. Thus in fact, $X_{n+1} \leq \bar{X}_{n+1}$ must hold, which proves (ii).

ad (iii): Let $X \in \mathcal{O}$, and let $Y \in \mathcal{O}^\perp$ be optimal. By (i), we may choose an optimal $\bar{Y} \in \mathcal{O}^\perp$ which is complementary to X:

$$\begin{aligned} X &= +\oplus\ldots\oplus\ 0\ldots0\ * \in \mathcal{O}, \\ \bar{Y} &= *\ 0\ \ldots\ 0\ \ominus\ldots\ominus - \in \mathcal{O}^\perp. \end{aligned}$$

Then obviously, $X \perp \bar{Y}$ implies $X_{n+1} = \bar{Y}_0$. But $\bar{Y}_0 = Y_0$ by (ii). This proves (iii).

ad (iv): Suppose $\mathcal{O}$ is feasible. Then $\mathcal{O}$ is unbounded if and only if there exists a $Z \in \mathcal{O}$ as below:

$$\exists Z = 0 \oplus \ldots \oplus + \in \mathcal{O}.$$

Hence, by MINTY's Lemma, this means that

$$\nexists Y = * \ominus \ldots \ominus - \in \mathcal{O}^{\perp}$$

which means that $\mathcal{O}^{\perp}$ is infeasible.

□

Now we are ready to prove the main theorem of linear programming duality, ensuring the existence of optimal solutions:

Theorem 6.10 *Let $\mathcal{O}$ and $\mathcal{O}^{\perp}$ be a dual pair of oriented matroids on $E = \{0, \ldots, n+1\}$, and consider the corresponding pair of dual oriented matroid LPs:*

$$X = + \oplus \ldots \oplus \uparrow \in \mathcal{O} \quad \textit{and} \quad Y = \downarrow \ominus \ldots \ominus - \in \mathcal{O}^{\perp}.$$

If both of these problems are feasible, then both have optimal solutions. Thus an oriented matroid LP is either infeasible, unbounded or has optimal solutions.

Proof. The proof is by induction on n. For $n = 0$, there is nothing to show (in fact, any two sign vectors $X \in \mathcal{O}$ and $Y \in \mathcal{O}^{\perp}$ are complementary in this case). Thus assume that $n \geq 1$ and assume that the two dual problems

$$\begin{array}{lll} (1) & + \oplus \oplus \ldots \oplus \uparrow \in \mathcal{O} & \text{and} \\ (2) & \downarrow \ominus \ominus \ldots \ominus - \in \mathcal{O}^{\perp} & \end{array}$$

are feasible. We consider three cases.

Case 1: Every feasible $X \in \mathcal{O}$ has $X_1 = +$.
In this case, consider the two smaller problems

$$\begin{array}{lll} (3) & + * \oplus \ldots \oplus \uparrow \in \mathcal{O} & \text{and} \\ (4) & \downarrow 0 \ominus \ldots \ominus - \in \mathcal{O}^{\perp} & . \end{array}$$

Problem (3) is obviously feasible because (1) is feasible. Furthermore, if U is any feasible solution for (3), then U_1 must be nonnegative, otherwise eliminating coordinate 1 between U and any feasible solution X for (1), fixing coordinate 0, would result in a feasible solution for (1) contradicting our assumption of Case 1. Hence any feasible solution of (3) is in fact a feasible solution of (1). From this it is immediate that any pair of complementary optimal solutions for (3) and (4) yields a pair of complementary optimal solutions for (1) and (2).

By induction, what we are left to show is that (4) is feasible. To see this, note the following:

$$\begin{array}{llcl}
\text{Feasibility of } \mathcal{O} & \Rightarrow \exists X & = & + \oplus \oplus \ldots \oplus * \in \mathcal{O} \\
\text{Case 1} & \Rightarrow \not\exists Z & = & + 0 \oplus \ldots \oplus * \in \mathcal{O} \\
\text{Minty's Lemma} & \Rightarrow \exists U & = & - * \ominus \ldots \ominus 0 \in \mathcal{O}^{\perp} \\
U \perp X & \Rightarrow \exists U & = & - + \ominus \ldots \ominus 0 \in \mathcal{O}^{\perp} \\
\text{Feasibility of } \mathcal{O}^{\perp} & \Rightarrow \exists Y & = & * \ominus \ominus \ldots \ominus - \in \mathcal{O}^{\perp}
\end{array}$$

If $Y \in \mathcal{O}^{\perp}$ as above has component 1 equal to zero, then Y is feasible for (4). If component 1 of Y is negative, then eliminating component 1 between U and Y, fixing component $n+1$ results in a sign vector showing that (4) is indeed feasible. Hence case 1 is settled.

Case 2: Every feasible $Y \in \mathcal{O}^{\perp}$ has $Y_1 = -$.
This case can be treated in the very same way as Case 1.

Case 3: There are feasible $X \in \mathcal{O}$ and $Y \in \mathcal{O}^{\perp}$ with $X_1 = Y_1 = 0$.
In this case we consider the following smaller LPs:

$$\begin{array}{ll}
(5) & + 0 \oplus \ldots \oplus \uparrow \in \mathcal{O} \\
(6) & \downarrow * \ominus \ldots \ominus - \in \mathcal{O}^{\perp} \\
(7) & + * \oplus \ldots \oplus \uparrow \in \mathcal{O} \\
(8) & \downarrow 0 \ominus \ldots \ominus - \in \mathcal{O}^{\perp}
\end{array}$$

Note that (5) and (6), resp. (7) and (8) are dual pairs of smaller sized LPs. Furthermore, due to our assumption of Case 3, each of these four problems is feasible. Hence, by induction, there exist optimal complementary solutions $U \in \mathcal{O}$, $V \in \mathcal{O}^{\perp}$ for (5) and (6), and $U' \in \mathcal{O}$, $V' \in \mathcal{O}^{\perp}$ for (7) and (8). We claim that either (U, V) or (U', V') is a pair of optimal solutions for (1) and (2). Suppose not — then V and U' must violate the sign restrictions in the first component, i.e. we must have, say,

$$\begin{array}{ll}
U = & + 0 + \ldots + + \ldots + 0 \ldots 0 0 \ldots 0 * \in \mathcal{O} \\
V = & * + 0 \ldots 0 0 \ldots 0 \ominus \ldots \ominus \ominus \ldots \ominus - \in \mathcal{O}^{\perp} \\
U' = & + - 0 \ldots 0 + \ldots + + \ldots + 0 \ldots 0 * \in \mathcal{O} \\
V' = & * 0 \ominus \ldots \ominus 0 \ldots 0 0 \ldots 0 \ominus \ldots \ominus - \in \mathcal{O}^{\perp}
\end{array}$$

Approximating U and $-U'$ on the first component yields a sign vector

$$Z = 0 + + \ldots + * \ldots * - \ldots - 0 \ldots 0 * \in \mathcal{O}$$

Now $Z \perp V$ implies that $Z_{n+1} = +$. But then Z is improving with respect to U', contradicting our assumption that U' is an optimal solution of (7).

□

As mentioned earlier already, the abstract concept of oriented matroid programs allows to treat linear programming duality in a very concise and elegant way. However, the results as stated above may nevertheless appear somewhat strange or even crude to those who ever learned about linear programming duality before. Thus let us derive a more familiar version of LP duality in the traditional setting of "ordinary" linear programs. To do this, we simply have to look at what the subspace L and $L^\perp$, corresponding to $\mathcal{O}$ and $\mathcal{O}^\perp$ resp., are.

Consider a dual pair of LPs, and — as it is done in most textbooks on linear programming — assume, that the primal program is already in standard form:

$$\begin{aligned} \max\ & c^T x \\ Ax &= b \\ x &\geq 0. \end{aligned} \tag{6.13}$$

The dual of this may be formulated as

$$\begin{aligned} \min\ & b^T u \\ u^T A &\geq c. \end{aligned} \tag{6.14}$$

These correspond to (6.11) and (6.12), resp., with

$$L = ker \begin{pmatrix} 0 & c^T & -1 \\ -b & A & 0 \end{pmatrix} \text{ and } L^\perp = im \begin{pmatrix} 0 & c^T & -1 \\ -b & A & 0 \end{pmatrix}^T.$$

Let x and u denote feasible solutions of (6.13) and (6.14), resp. Then

$$\begin{aligned} x' &= (1, x^T, c^T x) && \text{is feasible in } L \\ \textit{and}\quad y' &= (1, -u^T) \begin{pmatrix} 0 & c^T & -1 \\ -b & A & 0 \end{pmatrix} && \text{is feasible in } L^\perp \end{aligned}$$

(and any feasible $x' \in L$ and $y' \in L^\perp$ can be obtained in this way).

Now, if x' and y' are complementary, this means that $x'_i \cdot y'_i = 0$ for $i = 1, \ldots, n$. Since $x_i \geq 0$ and $y_i \leq 0$ $(i = 1, \ldots, n)$, this is equivalent to

$$\sum_{i=1}^{n} x'_i y'_i = 0.$$

In terms of x and u, this means that

$$0 = \sum_{i=1}^{n} x_i (c_i - u^T A_{.i}) = x^T (c - A^T u).$$

In this case, x and u are called **complementary**. It is obvious now, how to translate the above duality results into the traditional setting of ordinary LPs. In particular, we get

Corollary 6.11 (Strong Complementary Slackness, Traditional Version) *Either both (6.13) and (6.14) are infeasible or both have optimal solutions or one of them is unbounded and the other is infeasible. Given a primal feasible solution x (dual feasible solution u), this is optimal if and only if there exists a complementary dual (primal) solution. Hence, in particular, x and y are optimal if and only if $c^T x = b^T u$.*

□

6.3 Network Programming

Let us consider a special class of problems where linear duality is an even more direct application of MINTY's Lemma. This class of problems is the socalled class of network flow problems to be introduced below. In electrical networks, linear duality reduces to what is known as the duality between current and voltage. As we will see, this is precisely the same as MINTY's Lemma. Thus, networks provide a more intuitive understanding of the rather technical MINTY-Lemma.

Consider a graph $G = (V,E)$ with two distinguished nodes s and t, which we call **source** and **sink**, resp. Suppose, e.g., that the edges of G are pipelines and that we want to transport as much oil per time as possible from s to t, subject to some restrictions concerning the throughput capacity of the edges (e.g. the diameter of the pipelines). A further condition we impose is, that at any vertex r different from s and t no oil is lost or added, i.e. the amount of the oil "entering" r equals the amount of oil "leaving" r. If x_e denotes the amount of oil which is (per unit of time) transported through the edge $e \in E$, this can be written as follows:

$$\textstyle\sum_{e \in \delta^+(r)} x_e - \sum_{e \in \delta^-(r)} x_e = 0. \qquad (6.15)$$

Definition 6.12 *Let $G = (V,E)$ be a digraph and let s and t be two nodes of G. Then a vector $x \in \mathbb{K}^E$ is called a* **flow** *(from s to t) if (6.15) holds for every node $r \in V \setminus \{s,t\}$. Furthermore, if we are given two* **capacity functions** *$l : E \to \mathbb{K}$ and $u : E \to \mathbb{K}$, then a flow x is called* **feasible**, *provided $l \le x \le u$ (in each component $e \in E$).*

Given a flow x, we may define its **value** to be the net amount of flow leaving s, i.e.

$$v(x) := \sum_{e\in\delta^+(s)} x_e - \sum_{e\in\delta^-(s)} x_e .$$

This is the same as the net amount of flow entering t, since no flow is added or lost in any node $r \neq s, t$. (Formally, this can be seen by adding up the net amounts of flow in all vertices. The sum equals 0 because every edge appears twice.)

The max flow problem now is to find a feasible flow whose value is maximal. This is obviously a linear programming problem. In fact, the set of feasible flows x is a polyhedron and the value $v(x)$ is a linear objective function. Thus in principle, we could discuss linear duality by reformulating the max flow problem as an LP in standard form, computing its dual and applying the results of section 6.2. However, let us choose a more direct approach here, because this turns out to be more instructive.

Suppose we are given any feasible flow, and let us try to find out whether it is optimal or, in case it is not, find a better solution. Thus let $x \in \mathbb{K}^E$ be a feasible flow. It is natural to try to improve x in the following way: Try to find a path p from s to t such that the current flow can be increased by some small amout $\varepsilon > 0$ in every forward edge of p and decreased (by the same small amount $\varepsilon > 0$) in every backward edge. Then obviously, the resulting flow, formally defined by $x' = x + \varepsilon p$, where p is understood as an incidence vector, will still be feasible. Formally, let us define

$$\begin{aligned} R &:= \{e \in E \mid l_e = x_e < u_e\}, \\ G &:= \{e \in E \mid l_e < x_e = u_e\}, \\ B &:= \{e \in E \mid l_e < x_e < u_e\}, \\ W &:= \{e \in E \mid l_e = x_e = u_e\}. \end{aligned}$$

Then we are looking for a path p which, considered as an incidence vector, satisfies

a) $p_R \geq 0\ ,\ p_G \leq 0\ ,\ p_B = *\ ,\ p_W = 0.$

Recall from Section 2.2 how we solved the similar (slightly simpler) problem of finding a directed path from s to t in a graph by successively constructing the set $S \subseteq V$ of nodes that can be reached from s by a directed path. This extends in an obvious way to the above problem of finding a path from s to t as in a). In fact, one may successively construct the set $S \subseteq V$ of nodes that are reachable from

s along a path satisfying the constraints given in a). Then either S contains t (and hence an $s-t$ path as in a) is discovered) or we get an $s-t$ cut given by the partition $V := S\dot{\cup}T$ such that all edges leaving S and entering T are contained in $G \cup W$ and all edges leaving T and entering S are contained in $R \cup W$. Thus, the incidence vector y corresponding to this cut satisfies

b) $y_R \leq 0\ ,\ y_G \geq 0\ ,\ y_B = 0\ ,\ y_W = *.$

In this case, obviously no $s-t$ path p as in a) above can exist, and in fact, it is easy to show that our current flow x must be optimal already. To see this, consider the incidence vector $y \in \mathbb{K}^E$ of an arbitrary s–t-cut given by a partition $V = S\dot{\cup}T$ and define its **capacity** to be

$$c(y) := \sum_{e \in y^+} u_e - \sum_{e \in y^-} l_e.$$

Intuitively, this gives the "maximum possible net amount" of flow that can be shifted from S to T. In fact, the capacity of any s–t-cut y provides an upper bound on the value of a maximum flow. For, let $x \in \mathbb{K}^E$ be a feasible flow, and let $y \in \mathbb{K}^E$ be an s–t-cut with corresponding partition $V = S\dot{\cup}T$, then by the flow conservation condition 6.15, we get

$$\begin{aligned} v(x) &= \sum_{e \in \delta^+(s)} x_e - \sum_{e \in \delta^-(s)} x_e = \sum_{e \in \delta^+(S)} x_e - \sum_{e \in \delta^-(S)} x_e \\ &= \sum_{e \in y^+} x_e - \sum_{e \in y^-} x_e \leq \sum_{e \in y^+} u_e - \sum_{e \in y^-} l_e = c(y). \end{aligned}$$

However, in case y is a special s–t-cut, satisfying condition b) as above, then it is immediate from the definition of R, G, B and W that the above inequality is satisfied with equality, hence x is an optimal solution. We have proved:

Theorem 6.13 (Max Flow–Min Cut Theorem) *The maximum value of a feasible s–t-flow equals the minimum capacity of an s–t-cut.*

□

Note that (from a strictly formal point of view) this result is not really a Linear Programming duality statement, because it relates the optimal value of an LP problem to the minimum of a capacity function c, defined on a finite set (the set of s–t-cuts $y \in \mathbb{K}^E$) rather than a polyhedron. Nonetheless, of course, Linear Programming duality is the general idea behind the above theorem and, as we have seen above,

MINTY's Lemma is essentially all what makes up its proof. As we mentioned already at the beginning of this section, one can also derive Theorem 6.13 by starting with the LP formulation of the max flow problem, constructing its dual and applying the Linear Programming duality results from Section 6.2. The reader is invited to do so and find out the LP associated to the min cut problem this way.

6.4 Further Reading

As we have seen, linear programming duality generalizes very naturally to the abstract setting of oriented matroids, or oriented matroid programs. A natural question that comes up next, is whether known algorithms for solving (ordinary) linear programs such as the Simplex Algorithm, can be translated and applied to solve abstract oriented matroid programming problems. This question has been studied, for example, in

R.G. BLAND *A Combinatorial Abstraction of Linear Programming*, **Journal of Combinatorial Theory B 23 (1977), pp. 33–57,**

K. FUKUDA *Oriented Matroid Programming*, **Thesis, University of Waterloo, Canada (1981), supervised by J. Edmonds.**

We do not want to go into that here. (After all, we did not even introduce the Simplex Algorithm for ordinary LPs.) However, it is quite easy to understand intuitively, why abstract linear programs may be more difficult to solve than ordinary ones. Basically, as indicated already in Section 6.2, the problem is that, given two feasible solutions, say X and X' of an oriented matroid program, we cannot tell whether X is "better" than X' or not, in case both X and X' have objective function coordinate equal to "+", say. In the ordinary case we can ensure any algorithm to make an actual, measurable progress in each step, by simply watching the objective function value grow. In the abstract setting, however, if we follow some strategy for going from one feasible solution to another one, the situation is different. And in fact, as turned out by investigations of Bland, Edmonds and Fukuda, a natural analogue of the Simplex Algorithm, applied to an oriented matroid program, may happen to produce a sequence of solutions step by step — and finally return to where it started from, i.e. run into a cycle and loop forever.

Such Problems may be overcome by restricting the class of oriented matroids to a smaller one, containing "nice" ones exclusively, as has been done by Jack Edmonds introducing the socalled BOMs

(Bland Oriented Matroids). They are further investigated by Fukuda [77] and Mandel [126]. Actually, they restrict themselves to oriented matroids satisfying an abstract analogue of Euclid's postulate (for every hyperplane H and every point $p \notin H$ there exists a hyperplane H' through p parallel to H). An even more restricted class of oriented matroids is studied in Bachem and Kern [6].

Another approach to overcome the above mentioned cycling problems consists in developing new methods (other than the Simplex Method) for solving Linear Programming problems. This has been achieved independently by T. Terlaky and Zh. Wang, cf.

T. TERLAKY *A Finite Crisscross Method for Oriented Matroids*, Journal of Combinatorial Theory, Series B 42 (1987), pp. 319–327,

ZH. WANG *A Finite Conformal-Elimination-Free Algorithm for Oriented Matroid Programming*, Chinese Annals of Mathematics, 8B(1) (19875).

Those who want to learn more about general "ordinary" LP techniques may consult any of the linear programming textbooks mentioned at the end of Chapter 3. Combinatorial problems such as max flow and other applications can be found in

C.H. PAPADIMITRIOU, K. STEIGLITZ *Combinatorial Optimization*, Prentice Hall Inc. (1982).

Finally, we would also like to mention the contribution of M. Todd to abstract LP duality and related subjects:

M. TODD *Linear and Quadratic Programming in Oriented Matroids*, Journal of Combinatorial Theory B 39 (1985), 105–133,

M. TODD *Complementarity in Oriented Matroids*, SIAM Journal of Algebraic and Discrete Mathematics 5(4) (1984), pp. 426–445.

Chapter 7

Basic Facts in Polyhedral Theory

As we have seen, oriented matroids provide a natural way to study linear programming in an abstract setting. A major second field of "application" is to study the structure of polyhedra in the general framework of oriented matroids. This will be our main object in the following. Our investigation starts with the present chapter, introducing some basic notions and results from polyhedral theory. In particular, we will prove the two representation theorems, i.e. MINKOWSKI's Theorem, which states that every polyhedron $P \subseteq \mathbb{K}^n$ can be represented as $P = conv\, V + cone\, E$ for two finite sets $V, E \subseteq \mathbb{K}^n$, and WEYL's Theorem, which states that every set $P \subseteq \mathbb{K}^n$ represented this way actually is a polyhedron. Furthermore, we will introduce the concept of polarity by associating a **polar** pair of polyhedral cones $C = P(B,0)$ and $C^P = cone\, B^T$ to every subspace $L = im\, B \subseteq \mathbb{K}^m$ (cf. Section 7.2).

We would like to stress, however, in advance that all these results are by no means anything new. They are nothing but **polyhedral interpretations** of duality between vectorspaces — in the same sense as, e.g., Corollary 4.2 is a polyhedral interpretation of Theorem 4.6. Thus, e.g., MINKOWSKI's Theorem is a polyhedral interpretation of the Composition Theorem, polarity is a polyhedral interpretation of duality and WEYL's Theorem is in a sense the sum of both. Essentially, this interpretation is done by means of "delinearization" (cf. Section 4.4), that is, by applying a linear transformation: If we are given a matrix $B \in \mathbb{K}^{m \times n}$, then everything that happens in $L = im\, B \leq \mathbb{K}^m$ corresponds to something happening in $\mathbb{K}^n$ and hence may be interpreted in terms of $C = P(B,0)$. Hence we may study subspaces instead of polyhedral cones and vice versa. Both aspects have their advantages: Duality Theorems are usually easier to understand intuitively, if we interprete them in terms of polyhedral cones. On the other hand, proofs become clearer and more elegant in the vectorspace setting.

The most important relation between $L = im\, B$ and $C = P(B,0)$ is actually a relation between the oriented matroid $\mathcal{O} = \sigma(L)$, and the socalled "face lattice" of C, which is introduced in Section 7.5. In fact, the face lattices of both $C = P(B,0)$ and its polar $C^P = cone\, B^T$ will turn out to be geometrical interpretations of the partial order $\preceq$, defined on $\mathcal{O} = \sigma(L)$. There is no such straightforward geometrical interpretation for general oriented matroids which do not arise from subspaces $L \leq \mathbb{K}^m$. Therefore, throughout this chapter, everything will take place in $\mathbb{K}^m$ and $\mathbb{K}^n$. (But wait for Chapter 9!)

7.1 MINKOWSKI's Theorem

As polyhedral cones are much easier to handle than general polyhedra, we will prove MINKOWSKI's Theorem for cones first. The corresponding result for general polyhedra may then be derived by means of dehomogenization. As we will see, the "homogenized" version of MINKOWSKI's Theorem we are going to present below is quite an easy consequence of the Composition Theorem — in fact, the former is just a polyhedral interpretation of the latter.

Definition 7.1 *Let $C = P(B,0)$ be a polyhedral cone in $\mathbb{K}^n$. Then $x \in C$ is called* **elementary** *if the corresponding vector $y = Bx \leq 0$ is an elementary vector in $L = im\, B$.*

Note that, at the first glance, this definition is somewhat unsatisfactory, since whether or not a given $x \in C$ is elementary, seems to depend on the particular representation of C by means of the matrix B. However, this is not true, as can be seen from Lemma 7.4 and Lemma 7.5 below.

Example 7.2

1. *If $C = lineal\, C$ is a subspace of $\mathbb{K}^n$ then C does not have any elementary vectors, since $y = Bx = 0$ for every $x \in C$ (cf. Section 1.4).*
2. *If $C = \mathbb{K}_+ \times \mathbb{K} = \{x \in \mathbb{K}^2 \mid x_1 \geq 0\}$ then every $x \in C$ with $x_1 > 0$ is elementary.*
3. *If $C = \mathbb{K}_+^2$ then $x = \binom{1}{0}$ and $x' = \binom{0}{1}$ are, up to scaling, all elementary vectors of C.*

This last example indicates that elementary vectors are in a sense "extreme", in case the lineality space equals 0.

Definition 7.3 *A polyhedral cone with lineality space equal to 0 is called* **pointed**. *An element $x \neq 0$ of a polyhedral cone C is called* **extreme** *if the following holds: Whenever x is a proper convex combination of $x', x'' \in C$, then $x', x'' \in \mathbb{K}_+x = \{\lambda x \mid \lambda \in \mathbb{K}_+\}$. If $x \in C$ is extreme, then $\mathbb{K}_+x$ is called an* **extreme ray** *of C.*

There is a pointed cone associated to every polyhedral cone in the following way:

Lemma 7.4 *Let C be a polyhedral cone in $\mathbb{K}^n$. Then*

$$C_0 := C \cap (\mathit{lineal}\, C)^{\perp}$$

is a pointed polyhedral cone. Moreover, $C = C_0 + \mathit{lineal}\, C$.

Proof. C_0, being an intersection of two polyhedral cones, is of course a polyhedral cone again. Its lineality space is equal to $\mathit{lineal}\, C \cap (\mathit{lineal}\, C)^{\perp} = 0$. This proves the first claim. Furthermore, if $x \in C$ then $x = u + w$ with $u \in \mathit{lineal}\, C$ and $w \in (\mathit{lineal}\, C)^{\perp}$. By definition of the lineality space, $w = x - u \in C$, hence $w \in C_0$. Thus $x = u + w \in C_0 + \mathit{lineal}\, C$. This shows that $C \subseteq C_0 + \mathit{lineal}\, C$. The converse inclusion is trivial.

Note that the decomposition $x = w + u$ is unique, since $u \in \mathit{lineal}\, C$ and $w \in (\mathit{lineal}\, C)^{\perp}$.

□

Lemma 7.5 *Let $C \neq 0$ be a polyhedral cone. Then C has an extreme ray if and only if C is pointed. In this case, $x \in C$ is extreme if and only if it is elementary. Hence a pointed polyhedral cone has only finitely many extreme rays.*

Proof. Let $C = P(B, 0)$. If C is not pointed, then $\ker B = \mathit{lineal}\, C \neq 0$. Choose $0 \neq u \in \mathit{lineal}\, C$. Then every $x \in C$ is a proper convex combination of $x'_\lambda := x - \lambda u$ and $x''_\lambda = x + \lambda u, \lambda > 0$. For sufficiently large $\lambda > 0$, either x'_λ or x''_λ is not in $\mathbb{K}_+x$. Hence C cannot have extreme rays. Now let C be pointed, i.e. $\ker B = 0$. This means that the mapping: $x \longrightarrow y = Bx$ is $1-1$, implying that $x \in C$ is a proper convex combination of $x', x'' \in C$ if and only if $y = Bx$ is a proper convex combination of $y' = Bx'$ and $y'' = Bx''$. From this one concludes that x is extreme if and only if it is elementary. Thus in particular, C does have extreme vectors in this case.

□

After these preliminaries, let us turn to MINKOWSKI's Theorem:

Theorem 7.6 (MINKOWSKI) *Every pointed polyhedral cone is the conic hull of its (finitely many) extreme rays.*

Proof. Let $C = P(B, 0)$ be pointed, and let $E \subseteq C$ be such that $\{\mathbb{K}_+e \mid e \in E\}$ is the set of the extreme rays of C. We claim that $C = cone\, E$.

"$\supseteq$" is trivial.

"$\subseteq$": Let $x \in C$, and let $y = Bx$. Then $y \leq 0$ is the conformal sum of elementary vectors $y^i \in im\, B$, i.e. $y = y^1 + \cdots + y^k$, $k \geq 0$. Let x^i be corresponding elements of C, i.e. $y^i = Bx^i \;\; (i = 1, \cdots, k)$. Then the x^i's are elementary, and since C is pointed, they are extreme. Since C is pointed, the mapping $x \rightarrow y = Bx$ is $1 - 1$, implying that $x = \sum x^i$. This shows that $x \in cone\, E$.

□

Corollary 7.7 *Every polyhedral cone C can be written as a conic hull of finitely many vectors. More precisely,*

$$C = cone\, E + lineal\, C$$

for a finite set $E \subseteq \mathbb{K}^n$.

Proof. This follows from MINKOWSKI's Theorem and Lemma 7.4. Note that this actually *is* a conic hull representation since $lineal\, C = cone(E' \cup -E')$ where E' is any basis of $lineal\, C$.

□

7.2 Polarity

Duality between vectorspaces is not a really difficult theory, and the Duality Theorems it provides are not very hard to understand, too (once you are used to work with them). From a geometrical point, however, our understanding is still very poor. If we try to imagine a dual pair of subspaces $(L, L^\perp)$ in $\mathbb{K}^n$ then most of us will have to content themselves with $n = 3$. (There exist people, who claim that they have no problems with imagining higherdimensional spaces. In case, they are mathematicians, they are usually called "geometers".) The loss of intuition, due to the above mentioned widespread inability of human race, may sometimes be compensated (to some extent) by associating low dimensional polyhedral cones to high dimensional subspaces. More precisely, to every subspace $L = im\, B \leq \mathbb{K}^m$, there are two cones associated in a natural way, namely $C = P(B,0)$ and $C^P = cone\, B^T$. The former is polyhedral by definition and the latter will be proven to be polyhedral below. An obvious relation between these two is the following: If $x \in C$ and $y \in C^P$ then $x^T y \leq 0$.

Definition 7.8 *Let $S \subseteq \mathbb{K}^n$. Then*

$$S^P := \{y \in \mathbb{K}^n \mid y^T s \leq 0 \quad \forall s \in S\}$$

is called the **polar cone** *of S, or simply the* **polar** *of S. In other words, $y \in S^P$ if and only if $y^T s \leq 0$ is valid for S, i.e. S^P can be regarded as a cone of "valid inequalities".*

Lemma 7.9

(i) *For $S \subseteq \mathbb{K}^n$, $S^P = cone\, S^P = (cone\, S)^P$ and $S^{PP} \supseteq S$.*

(ii) *For $S, T \subseteq \mathbb{K}^n$, $(S \cup T)^P = S^P \cap T^P$.*

(iii) *For a subspace $L \leq \mathbb{K}^n$, $L^P = L^\perp$. More generally, if $S \subseteq \mathbb{K}^n$ then $S^\perp = S^P \cap (-S)^P$.*

Proof. This is left to the reader as a straightforward exercise.

□

The following turns out to be essentially a restatement, or a polyhedral interpretation of FARKAS' Lemma:

Proposition 7.10 *The polar of $C = P(B,0)$ is $C^P = cone\, B^T$ and vice versa. Hence, in particular, $C = C^{PP}$ for every polyhedral cone.*

Proof. Let E denote the set of rows of B. Then by Lemma 7.9 (i), $cone(B^T)^P = cone(E)^P = E^P = P(B,0)$, i.e. $P(B,0)$ is the polar of $cone\, B^T$. Next, let us show that $cone\, B^T = P(B,0)^P$.

"$\subseteq$" is trivial.

"$\supseteq$": Let $x \in \mathbb{K}^n$ and $x \notin cone\, B^T = cone\, E$. By Theorem 4.9, x can be separated from E, i.e. there exists $u \in \mathbb{K}^n$ such that $Bu \leq 0$ and $x^T u > 0$. Hence $x^T u > 0$ for some $u \in P(B,0)$, i.e. $x \notin P(B,0)^P$.

□

Corollary 7.11 *If $C \subseteq \mathbb{K}^n$ is a polyhedral cone, then $lin\, C^P = (lineal\, C)^\perp$.*

Proof. Let $C = P(B,0)$. Then $C^P = cone\, B^T$, hence $lin\, C^P = lin\, B^T = im\, B^T$ and $lineal\, C = ker B$.

□

Note that, by Proposition 7.10, FARKAS' Lemma may be restated as follows: For every $y \in \mathbb{K}^n, y^T x \leq 0$ is valid for $C = P(B,0)$ if and only if $y \in C^P = cone\, B^T$. This may help us to get a better understanding of FARKAS' Lemma sometimes: Given $L = im\, B$ with $B \in \mathbb{K}^{10\times 3}$, say, then the "linear" version of FARKAS' Lemma (Theorem 4.6) is a statement about L and $L^\perp$ being subspaces of $\mathbb{K}^{10}$, while the above version relates C and C^P, which are both contained in $\mathbb{K}^3$. This "polyhedral" point of view of a subspace $L \leq \mathbb{K}^n$ (or, more generally, an oriented matroid $\mathcal{O}$) will be emphasized throughout in what follows.

An easy consequence of MINKOWSKI's Theorem is the following:

Corollary 7.12 *If $C \subseteq \mathbb{K}^n$ is a polyhedral cone, so is its polar.*

Proof. By Corollary 7.6, C can be written as $cone\, E$ for some finite set $E \subseteq \mathbb{K}^n$. Let B denote the matrix whose rows are the vectors in E. Then $C = cone\, B^T$ and hence $C^P = P(B,0)$ is a polyhedral cone.

□

A further immediate consequence is a homogenized version of WEYL's Theorem, which we state separately:

Theorem 7.13 (WEYL) *For any finite set $E \subseteq \mathbb{K}^n$, $cone\, E$ is a polyhedral cone.*

Proof. By Proposition 7.10, $cone\, E$ is the polar of a polyhedral cone, and by Corollary 7.12, this is a polyhedral cone again.

□

Note that we obtained a proof of WEYL's Theorem by combining Proposition 7.10 and Corollary 7.12. These, in turn, are essentally due to FARKAS' Lemma and MINKOWSKI's Theorem, resp. On the other hand, WEYL's Theorem implies both FARKAS' Lemma (in fact it strengthens Corollary 4.12) and MINKOWSKI's Theorem. The latter may be seen as follows: As WEYL's Theorem implies FARKAS' Lemma, it implies Proposition 7.10, hence, in particular, $C = C^{PP}$ for every polyhedral cone $C = P(B,0)$. Since $C^P = cone\, B^T$ is a polyhedral cone by WEYL's Theorem, we may write $C^P = P(A,0)$ for some matrix A, hence $C = C^{PP} = P(A,0)^P = cone\, A^T$, which is MINKOWSKI's Theorem. Summarizing, we might say that WEYL equals FARKAS plus MINKOWSKI. Of course, this statement does not mean anything, since all three are valid in $\mathbb{K}^n$. However, this can be made precise in a more general setting, in which WEYL's Theorem is equivalent to FARKAS' Lemma plus MINKOWSKI's Theorem, but neither of them holds in advance. (Essentally this more general setting arises by replacing $\mathbb{K}$ by a more general structure, e.g. a ring or a semiring, cf. [38].)

7.3 Faces of Polyhedral Cones

In Section 7.1 we introduced the notion of extreme rays: If $C \subseteq \mathbb{K}^n$ is a polyhedral cone and $F = \mathbb{K}_+e$ is an extreme ray of C then F is a polyhedral cone which may be considered as an "extreme subcone" of C. Generalizing this notion, we may define extreme subcones of C in the following way:

Definition 7.14 *Let $C \subseteq \mathbb{K}^n$ be a polyhedral cone. Then a polyhedral cone $F \subseteq C$ is called an* **extreme subcone** *or a* **face** *of C if the following holds: If $x \in F$ is a proper convex combination of $x', x'' \in C$ then $x', x'' \in F$. (Since F and C are cones, this amounts to say that $x = x' + x'' \in F$ implies $x', x'' \in F$.) A face F is called* **proper** *if $F \subset C$. A maximal proper face of C is also called a* **facet** *of C. The* **dimension** *of a face F is defined to be $\dim F := \dim(\text{lin } F)$.*

Example 7.15

(1) *Let $C = \mathbb{K}^2$. Then $F = \mathbb{K}_+ \times \mathbb{K} \subseteq C$ is a polyhedral cone, but not a face of C. In fact, the only face of C is C itself.*

(2) *Let $C = \mathbb{K}_+ \times \mathbb{K}$. Then the faces of C are $\{0\} \times \mathbb{K}$ and C. They have dimensions equal to 1 and 2,resp..*

(3) *Let $C = \mathbb{K}_+^2$. Then the faces of C are $\{0\}$, $\mathbb{K}_+\binom{1}{0}$, $\mathbb{K}_+\binom{0}{1}$, and C. Their dimensions are, resp., 0,1,1 and 2.*

The following two results are straightforward consequences of the definition of faces.

Lemma 7.16 *Let C be a polyhedral cone, and let $\mathcal{F}$ denote the set of faces of C. Define a relation "$\leq$" on $\mathcal{F}$ by setting $F \leq F'$ if and only if F is a face of F' or, equivalently, $F \subseteq F'$. Then "$\leq$" is a partial order on $\mathcal{F}$.*

Lemma 7.17 *If F and F' are two faces of C then $F \cap F'$ is also a face of C.*

Before we analyse faces in more detail, let us reduce the general case to pointed cones:

Lemma 7.18 *Let C be a polyhedral cone, and let C_0 be the corresponding pointed cone, i.e. $C_0 = C \cap lineal\, C^\perp$ (cf. Lemma 7.4). Then F_0 is a face of C_0 if and only if $F = F_0 + lineal\, C$ is a face of C. Moreover, $F_0 \to F_0 + lineal\, C$ yields a $1-1$ correspondence between the faces of C_0 and the faces of C.*

Proof. Recall from Lemma 7.4 that $C = C_0 + lineal\, C$ and every $x \in C$ uniquely determines $x_0 \in C_0$ and $x_l \in lineal\, C$ such that $x = x_0 + x_l$. Now let $F \subseteq C$, and let $F_0 := F \cap lineal\, C^\perp$ be the corresponding subset of C_0. Let $x \in F$ and $x', x'' \in C$, and let $x_0 \in F_0$ and $x_0', x_0'' \in C_0$ correspond to x, x' and x'', resp. Then by the uniqueness of x_0, x_0' and x_0'' it follows that $x = x' + x''$ if and only if $x_0 = x_0' + x_0''$. From this the first claim follows immediately. To prove the second claim, we have to show that every face of C can be written as $F_0 + lineal\, C$. Thus let F be a face of C, and let $F_0 = F \cap lineal\, C^\perp$. Then $F = F_0 + lineal\, F$, hence we have to show that $lineal\, F = lineal\, C$. "$\subseteq$" is trivial. To prove the converse inclusion, let $u \in lineal\, C$. If $x \in F$ then $u \in lineal\, C$ implies that for every $\lambda > 0$ both $x' = x + \lambda u$ and $x'' = x - \lambda u$ are in C. Since $x \in F$ is a proper convex combination of x', x'', we conclude that $x', x'' \in F$. Hence $x + \lambda u \in F$ for every $\lambda \in \mathbb{K}$, i.e. $u \in lineal\, F$.

□

The following slightly strengthens MINKOWSKI's Theorem:

Proposition 7.19 *Let C be a pointed polyhedral cone. Then every face of C is a conic hull of (finitely many) extreme rays of C.*

Proof. First note that every face F of C is pointed, since $lineal\, F \leq lineal\, C = 0$. By MINKOWSKI's Theorem $F = cone\, E$ where $\{\mathbb{K}_+ e \mid e \in E\}$ is the set of extreme rays of F. Therefore, it suffices to show that every extreme ray of F is also an extreme ray of C. But this is clear, since the relation "$\leq$" ("is a face of") is transitive, cf. Lemma 7.16.

□

Corollary 7.20 *A polyhedral cone has only finitely many faces.*

Proof. This follows immediately from Proposition 7.19, in case the polyhedral cone is pointed (since it has only finitely many extreme rays). The general case may then be derived from Lemma 7.18.

□

Let us finish this section with an alternative characterization of faces which is more or less a reformulation of the original definition:

Proposition 7.21 *Let C be a polyhedral cone. Then $F \subseteq C$ is a face of C if and only if $C \setminus F$ is convex and $F = C \cap lin\, F$.*

Proof.

"$\Leftarrow$": Let $F = C \cap lin\, F$ such that $C \setminus F$ is convex. Then $F \subseteq C$ is a polyhedral cone. Furthermore, let $x', x'' \in C$, and let $x = x' + x'' \in F$. Since $C \setminus F$ is convex at least one of x' and x'' must be in F. Say, $x' \in F$. But then $x'' = x - x' \in lin\, F$ and hence $x'' \in C \cap lin\, F = F$, too. This shows that F is a face.

"$\Rightarrow$": Let $F \subseteq C$ be a face. Then, in particular $C \setminus F$ is convex. Furthermore, let $x'' \in C \cap lin\, F$, say $x'' = x - x'$ for some $x, x' \in F$. Then $x = x' + x'' \in F$ implies that $x'' \in F$, too. Hence $C \cap lin\, F \subseteq F$. The converse inclusion is trivial.

□

7.4 Faces and Interior Points

Since a face of a polyhedral cone $C \subseteq \mathbb{K}^n$ is a polyhedral cone again, we know that it can be represented by an inequality system $Ax \leq 0$. The original definition of faces, however, does not tell us how to get the system $Ax \leq 0$ from a system $Bx \leq 0$ representing the original cone C. This section is to derive a characterization of faces, yielding an explicit description of faces of $C = P(B, 0)$ in terms of the system $Bx \leq 0$. The results will be derived using the following notion of "interior points":

Definition 7.22 *Let $P \subseteq \mathbb{K}^n$ be a polyhedron. Then $x \in P$ is called an* **interior point** *of P if x is in the interior of P, considered as a subset of the topological space $L = aff\, P$. (Where L is endowed with the topology induced from $\mathbb{K}^n$.) The set of interior points of P is denoted by $int\, P$. Thus if d denotes the dimension of L, i.e. $L = aff(x_0, x_1, \ldots, x_d)$ and $B = B_n \cap L$ denotes the d-dimensional unit ball in L, then $x \in int\, P$ is equivalent to $x + \varepsilon B \subseteq P$ for some $\varepsilon > 0$. Note that, by convention, if $P = \{x\}$ consists of a single point, then $int\, P = P = \{x\}$. The boundary of a polyhedron P, denoted by ∂P, will always mean the boundary of P considered as a subset of $L = aff\, P$. Thus $\partial P = P \setminus int\, P$, since P is a closed subset of L.*

The definition has been given for general polyhedra $P \subseteq \mathbb{K}^n$ rather than just polyhedral cones only for further use in Section 7.8. For the time being, we will deal with polyhedral cones only. Note that if $C \subseteq \mathbb{K}^n$ is a cone, then $aff\, C$ can be replaced by $lin\, C$.

The above definition gives a topological characterization of interior points. There is an equivalent algebraic characterization which we are going to derive next.

Definition 7.23 *Let $C = P(B, 0)$ for some $B \in \mathbb{K}^{m \times n}$. If $x \in C$ then*

$$Q(x) := \{i \in \{1, \ldots, m\} \mid B_{i.}x = 0\}$$

is called the **equality set** *of x. More generally, if $S \subseteq C$, then*

$$Q(S) = \bigcap_{x \in S} Q(x)$$

is called the **equality set** *of S.*

Lemma 7.24 *Let $C = P(B, 0)$. Then $x \in C$ is an interior point of C if and only if $Q(x) = Q(C)$.*

Proof.

"$\Rightarrow$": Let $x \in int\, C$. Then $Q(x) \supseteq Q(C)$ by definition, thus we are left to show that the reverse inclusion holds, i.e. that $Q(x) \subseteq Q(x')$ for every $x' \in C$. Thus let $i \in Q(x)$ and $x' \in C$. Then $u = x - x' \in lin\, C$. Since $x \in int\, C$, this implies $x + \varepsilon u \in C$ for some $\varepsilon > 0$. Thus, in particular, $0 \geq B_{i.}(x + \varepsilon u) = \varepsilon B_{i.}u = -\varepsilon B_{i.}x'$. Hence $B_{i.}x' \geq 0$ and since $x' \in C$ this implies $i \in Q(x')$.

"$\Leftarrow$": Suppose $Q(x) = Q(C)$. Let u be an element of the unit ball in $lin\, C$, i.e. $u = x' - x''$ for some $x', x'' \in C$ and $\|u\| \leq 1$. Then $Q(x) = Q(C) \subseteq Q(x') \cap Q(x'')$. Thus, whenever we have $B_{i.}x = 0$ for some i then $B_{i.}u = 0$. Choosing $\varepsilon > 0$ small enough, we get $x + \varepsilon u \in C$ for every u in the unit ball in $lin\, C$. Thus $x \in int\, C$.

□

Corollary 7.25 *Let $C \subseteq \mathbb{K}^n$ be a polyhedral cone. Then $int\, C \neq \emptyset$.*

Proof. Write $C = P(B, 0)$. It follows from the definition of equality sets that for every $i \notin Q(C)$ there exists some $x^i \in C$ such that $i \notin Q(x^i)$, i.e. $B_{i.}x^i < 0$. Then obviously $\bar{x} = \sum_{i \notin Q(C)} x^i$ is an interior point of C. (Note that the empty sum corresponds to $\bar{x} = 0$ by convention.)

□

Now let us turn to the description of faces in terms of the matrix B. As a first step we prove the following description of linear hulls:

Lemma 7.26 *Let F be a face of $C = P(B, 0)$. Then*

$$lin\, F = \{u \in \mathbb{K}^n \mid B_{i.}u = 0 \quad \forall i \in Q(F)\}.$$

Proof. "$\subseteq$" is immediate from the definition of $Q(F)$.

"$\supseteq$": Let $u \in \mathbb{K}^n$ such that $B_{i.}u = 0 \, \forall i \in Q(F)$. Let $x \in int\, F$. Then $B(x+\varepsilon u) \leq 0$ for $\varepsilon > 0$ sufficiently small. Hence $x' := x+\varepsilon u \in C$. Similarily, $x'' = x - \varepsilon u \in C$ for $\varepsilon > 0$ sufficiently small. Since x is a proper convex combination of $x', x'' \in C$, and F is a face, we conclude that $x', x'' \in F$. Thus $u = \lambda(x' - x'') \in lin\, F$.

□

Now we are ready to prove our main theorem, describing faces in terms of B.

Theorem 7.27 *Let $C = P(B, 0)$ be a polyhedral cone. Then $F \subseteq C$ is a face of C if and only if*

$$F = C \cap \{u \in \mathbb{K}^n \mid B_{i.}u = 0 \quad \forall i \in Q(F)\}.$$

Proof. "$\Rightarrow$" is an immediate consequence of Proposition 7.21 and Lemma 7.26.

"$\Leftarrow$" is straightforward from the definition of faces. Let F be as assumed in the claim. Then $F \subseteq C$ is a polyhedral cone. Furthermore, suppose that $x = x' + x'' \in F$ for some $x', x'' \in C$, i.e. $Bx' \leq 0$ and $Bx'' \leq 0$. Then for every $i \in Q(F)$ we have $0 = B_{i.}x = B_{i.}x' + B_{i.}x'' \leq 0$, implying that $x', x'' \in F$. This shows that F is a face.

□

The above theorem is sometimes stated in a slightly different form:

Definition 7.28 *Let $C \subseteq \mathbb{K}^n$ be a polyhedral cone, and let $cx \leq 0$ be a valid inequality for C. Then $H = c^\perp$ is called a* **supporting hyperplane** *of C. (Note that $c = 0$ is not excluded.)*

Corollary 7.29 *$F \subseteq C$ is a face if and only if $F = C \cap H$ for some supporting hyperplane H of C.*

Proof.

"$\Rightarrow$": Let F be a face of C. Then

$$F = C \cap \{u \in \mathbb{K}^n \mid B_{i.}u = 0 \quad \forall i \in Q(F)\}.$$

Let $c := \sum_{i \in Q(F)} B_{i.}$, and $H := c^\perp$. Then H is a supporting hyperplane of C and $F = C \cap H$.

"$\Leftarrow$": Let $H = c^\perp$ be a supporting hyperplane of C such that $F = C \cap H$. Then $C \setminus F = \{x \in C \mid cx < 0\}$ is a convex set. Furthermore, $lin\, F = lin\, C \cap H$, hence $C \cap lin\, F = C \cap H = F$. Thus F is a face by Proposition 7.21.

□

7.5 The Canonical Map

The most important aspect of Theorem 7.27 is that it allows us to study the faces of $C = P(B,0)$ in terms of the linear space $im\, B$ or, more precisely, in terms of $\mathcal{O} = \sigma(im\, B)$. In fact, consider the map

$$\sigma \circ B : \mathbb{K}^n \to \mathcal{O}.$$

This will be called the **canonical map** from $\mathbb{K}^n$ to $\mathcal{O}$.

Proposition 7.30 *The canonical map induces a 1-1 correspondence between the faces of C and the set $\{Y \in \mathcal{O} \mid Y \leq 0\}$. If F is a face of C, the corresponding sign vector is given by $Y = \sigma \circ B(int\, F)$. If $Y \in \mathcal{O}, Y \leq 0$, the corresponding face F of C is given by $F = \{x \in \mathbb{K}^n \mid \sigma \circ B(x) \preceq Y\}$, i.e. $F = (\sigma \circ B)^{-1}[Y]$. This correspondence is such that $F \leq F'$ (cf. Lemma 7.16) if and only if $Y \preceq Y'$ for the corresponding sign vectors.*

Proof. By Theorem 7.27, F is a face of C if and only if

$$F = C \cap \{x \in \mathbb{K}^n \mid B_{i.}x = 0 \quad \forall i \in Q(F)\}.$$

Thus there is a 1-1 correspondence between faces F of C and their equality sets $Q(F)$. On the other hand, every equality set $Q(F)$ corresponds to a unique $Y \in \mathcal{O}$ such that $Y \leq 0$ and $Y^0 = Q(F)$.

□

Thus the $\preceq$ relation on $\mathcal{O}$ may be interpreted as "being a face of ...". Of course, so far, we have only considered a small part of $\mathcal{O}$, namely the set $\{Y \in \mathcal{O} \mid Y \leq 0\}$, i.e. the set $\{Y \in \mathcal{O} \mid Y \preceq (-,\ldots,-)\}$. But the results are easily extended to the whole set $\mathcal{O}$:

Definition 7.31 *A cone $C \subseteq \mathbb{K}^n$ is called* **definable** *from $Bx \leq 0$ if and only if*

$$C = \{x \in \mathbb{K}^n \mid \sigma \circ B(x) \preceq X\}$$

for some $X \in 2^{\pm E}$.

Alternatively, a cone is definable from $Bx \leq 0$ if its defining system of inequalities is obtained from $Bx \leq 0$ by replacing some of the inequalities "$\leq$" by "$\geq$" or "$=$".

Theorem 7.32 *The canonical map induces a 1-1 correspondence between cones definable from $Bx \leq 0$ and the sign vectors in $\mathcal{O}$. If F and F' are two cones definable from $Bx \leq 0$, and Y, Y' are the corresponding sign vectors then $F \leq F'$ if and only if $Y \preceq Y'$.*

Proof. This is a straightforward extension of Proposition 7.30, obtained by replacing some of the inequalities $B_{i.}x \leq 0$ by $B_{i.}x \geq 0$ or $B_{i.}x = 0$.

□

Theorem 7.32 provides, in a sense, a geometrical (or topological) realization of a linear OM $\mathcal{O} = \sigma(im\, B)$. The purely "combinatorial" structure of the poset $(\mathcal{O}, \preceq)$ is realized by a system of cones in $\mathbb{K}^n$ such that the $\preceq$ relation in $\mathcal{O}$ corresponds to the $\leq$ relation "is a face of" in the system of cones.

Definition 7.33 *Let $Bx \leq 0$ be given, and let $\mathcal{O} = \sigma(im\, B)$. Let $\mathcal{S}$ be the set of cones definable from $Bx \leq 0$. Then every $C \in \mathcal{S}$ is also called a* **cell**. *If $Y \in \mathcal{O}$, let $C_Y \subseteq \mathbb{K}^n$ denote the corresponding cell. (This notion will become clear in Chapter 9 where we will be concerned with the problem of finding a topological realization — in the above sense — for general oriented matroids.)*

Corollary 7.34 *Let $\mathcal{S}$ be as in the definition above. Then the canonical map induces a 1-1 correspondence between cells $C \in \mathcal{S}$ and elements $Y \in \mathcal{O}$. The correspondence $Y \leftrightarrow C_Y$ is given by*

$$C_Y = (\sigma \circ B)^{-1}[Y].$$

In particular, ∂C_Y corresponds to $[Y] \setminus \{Y\}$. (Recall from section 1.1 that $[Y] = \{X \mid X \preceq Y\}$ denotes the order ideal generated by Y.)

Proof. Only the last assertion needs to be proved. We know that $\sigma \circ B(int\, C_Y) = Y$. Hence $\sigma \circ B(\partial C_Y) \subseteq [Y] \setminus \{Y\}$. On the other hand, if $X \prec Y$ then C_X is a proper face of C_Y, hence it is contained in the boundary of C_Y.

□

If $n = 3$, we may try to illustrate the situation by means of dehomogenization, i.e. by intersecting the hyperplanes $B_{i.}x = 0$ with $\mathbb{K}^2 \times \{1\}$, say, to get a two dimensional configuration as sketched in Figure 7.1 below. (In this figure, arrows indicate the projections of $B_{i.}$ onto $\mathbb{K}^2 \times \{1\}$ and the cells are labeled by their corresponding sign vectors.)

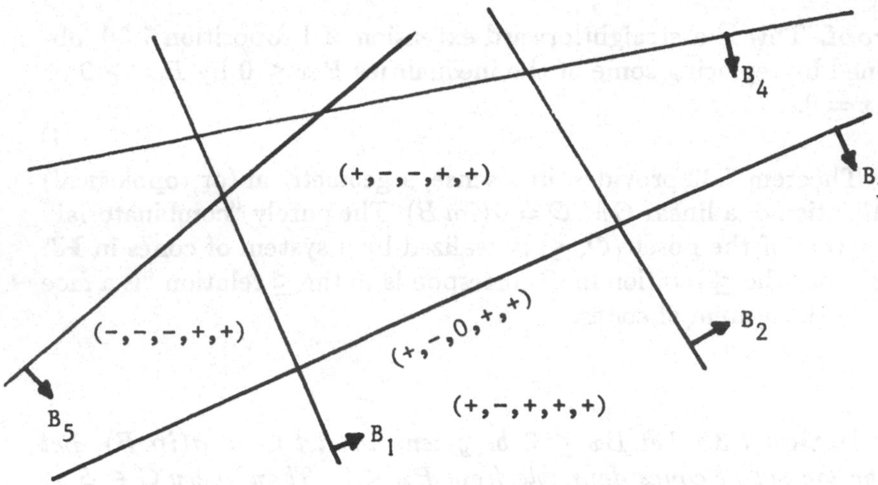

Figure 7.1

Definition 7.35 *Let $C \subseteq I\!K^n$ be a cone, and let $\mathcal{F}(C)$ denote the set of its faces. Then $\mathcal{F}(C)$, ordered by $\leq$, is called the* **face lattice** *of C. If $\mathcal{O}$ is an oriented matroid and $Y \in \mathcal{O}$ then $[Y]$, ordered by $\preceq$, is called the* **face lattice** *of* $[Y]$. (The notion of "lattices" will become clear in Section 7.6.)

As we noted already, the most important aspect of the canonical map is to provide a 1-1 correspondence between the face lattice of $[Y] \subseteq \mathcal{O}$, and the face lattice of its corresponding cone $C_Y \subseteq I\!K^n$. In fact, Theorem 7.32 states that these two posets are isomorphic. Hence we may study face lattices in $\mathcal{O} = \sigma(im\,B)$ instead of studying face lattices of polyhedral cones. There is no doubt that polyhedral theory, today, is almost exclusively concerned with investigating the structure of face lattices. One of the most important (and still unsolved) problems, for example, is the following: Given an arbitrary poset $(\mathcal{F}, \leq)$, is it the face lattice of a polyhedral cone? (We refer to the remarks in Section 8.5 for more about this problem.) Since this book is about oriented matroids, we will, of course be interested in studying face lattices in general oriented matroids rather than the special case of linear ones (corresponding to polyhedral cones). We can calm you, however, by saying that, up to now, no one has ever found a single "structural property" of polyhedral cone face lattices which is not owned also by general oriented matroid face lattices, exept for one thing: One can show (and this will be done in Section 7.7) that the class of face lattices of polyhedral cones is closed under antiisomorphisms, i.e. given any polyhedral cone C, there exists a polyhedral cone C', such that C and C' have antiisomorphic face lattices. This is **not** true for oriented matroid face lattices in general (cf. Section 7.9).

We will resume our study of oriented matroid face lattices in Chapter 8. Before, however, let us finish our introduction to basic polyhedral theory. Section 7.6 gives an a posteriori explanation for the notion of face *lattices*. Section 7.7 proves the above mentioned fact on antiisomorphisms. Section 7.8 shows how to apply dehomogenization in order to translate our results from polyhedral cones to general polyhedra.

7.6 Lattices

In this section we will introduce the concept of lattices, thereby explaining why face lattices of polyhedral cones are in fact "lattices". DEDEKIND has been the first to study lattices ("Dualgruppen") in depth. Since then, the notion of lattices has become a fundamental concept in mathematics. In fact, lattice theoretic concepts pervade the whole of modern algebra and have been proved useful in many other areas, including the foundations of set theory, general topology, geometry, and real analysis. The study of lattices is far beyond the scope of this book, and therefore, the interested reader is referred to standard textbooks on lattice theory, e.g. [20] and [92]. We will be concerned with finite lattices only, which do not make apparent the real power of lattice theory. They do provide, however, a convenient notation to work with.

Definition 7.36 *Let $(L, \leq)$ be a poset. If $x, y \in L$ then $z \in L$ is called an* **upper bound** *of x and y, provided $x \leq z$ and $y \leq z$. An upper bound z of x and y is called a* **least upper bound** *of x and y if $z \leq z'$ for every upper bound z' of x and y. Of course, if both z and z' are least upper bounds of x and y then $z \leq z'$ and $z' \leq z$, hence $z = z'$. Thus, if x and y do have a least upper bound, this is uniquely determined, and we denote it by $x \vee y$.*

Lower bounds *and* **greatest lower bounds** *are defined similarly. Thus z is a lower bound of x and y if $z \leq x$ and $z \leq y$. A lower bound z of x and y is called a greatest lower bound of x and y, provided $z \geq z'$ for every lower bound z' of x and y. If x and y have a greatest lower bound, this is uniquely determined, and we denote it by $x \wedge y$.*

$(L, \leq)$ is called a **lattice** *if any two elements $x, y \in L$ have a least upper bound $x \vee y$ and a greatest lower bound $x \wedge y$. $x \vee y$ and $x \wedge y$ are also called the supremum, resp. infimum of x and y in $(L, \leq)$. Obviously, $\wedge$ and $\vee$ are commutative and associative. In particular, if*

$x_1, \ldots, x_n \in L$ then

$$x_1 \wedge \ldots \wedge x_n \text{ and } x_1 \vee \ldots \vee x_n$$

are well defined elements of L, called the infimum and supremum of $x_1, \ldots, x_n$ in $(L, \leq)$.

Example 7.37

1. *Let $\mathbb{N} = \{1, 2, 3, \ldots\}$ and define a partial order $\leq$ on $\mathbb{N}$ by*

   $$x \leq y \Leftrightarrow x \text{ divides } y.$$

   Then $(\mathbb{N}, \leq)$ is a lattice with $\wedge$ and $\vee$ corresponding to "greatest common divisor" and "least common multiple", resp.
2. *Let $\mathcal{G}$ be a group, and L be the set of subgroups of $\mathcal{G}$. Define the relation $\leq$ on L by*

   $$H \leq H' \Leftrightarrow H \subseteq H'.$$

   Then $(L, \leq)$ is a lattice with $\wedge$ and $\vee$ corresponding to "intersection" and "generated subgroup", resp.
3. *Other examples, where $\leq$ is given by set-theoretic inclusion, are*
   - *the set of all subsets of a given set X,*
   - *the set of all subspaces of a given vectorspace X,*
   - *the set of all closed sets of a given topological space X.*

Let us state two further examples separately:

Proposition 7.38 *The face lattice of a polyhedral cone is a lattice.*

Proof. If C is a polyhedral cone then the order $\leq$ of $\mathcal{F}(C)$ is given by set-theoretic inclusion. Since $\mathcal{F}(C)$ is closed under taking intersections (cf. Lemma 7.17) we conclude that for any two faces $F', F'' \in \mathcal{F}(C)$

$$\begin{aligned} F' \wedge F'' &= F' \cap F'' \text{ and} \\ F' \vee F'' &= \bigcap \{F \in \mathcal{F}(C) \mid F \supseteq F' \cup F''\}. \end{aligned}$$

□

More generally, the following is true:

Proposition 7.39 *Let $\mathcal{O}$ be an oriented matroid, and let $Y \in \mathcal{O}$. Then $([Y], \preceq)$ is a lattice.*

Proof. Let Y' and Y'' be elements of $[Y]$. Since $[Y]$ is closed with respect to conformal sums, we conclude that

$$Y' \vee Y'' = Y' \circ Y'' \in [Y].$$

Furthermore, $Y' \wedge Y''$ is obviously equal to the conformal sum of all $X \in [Y]$ such that $X \preceq Y'$ and $X \preceq Y''$. Hence $Y' \wedge Y''$ also exists. □

Definition 7.40 *Let $(L, \leq)$ be a lattice, and let $L' \subseteq L$. Then L' (endowed with the order induced by $\leq$) is called a* **sublattice** *of L, provided*

$$x \wedge y \in L' \text{ and } x \vee y \in L' \quad \forall x, y \in L'.$$

If $(L, \leq)$ is any lattice then $(L, \geq)$, i.e. the lattice obtained by reversing the order $\leq$, is called the **dual** *of $(L, \leq)$. If the order $\leq$ is understood, we will often simply write L instead of $(L, \leq)$. The dual of L is then denoted by L^*. Obviously, L and L^* are antiisomorphic.*

As we will see in Section 7.7 below, the term "polar lattice" would be more appropriate for our purpose. However, the notion of "dual lattice" is well-established and therefore we did not feel authorized to change it. We warn the reader, however, that lattice duality as defined above is completely different from duality between subspaces L and $L^\perp$.

7.7 Face Lattices of Polars

In this section we will study the relationship between the face lattices of a polyhedral cone C and its polar C^P. The main result is as follows:

Theorem 7.41 *Let $C \subseteq \mathbb{K}^n$ be a polyhedral cone, and let C^P its polar. Then the face lattices $\mathcal{F}(C)$ and $\mathcal{F}(C^P)$ are antiisomorphic. The antiisomorphism $\varphi : \mathcal{F}(C) \to \mathcal{F}(C^P)$ is given by*

$$F \to C^P \cap F^\perp.$$

Proof. Let φ be as defined in the claim. We will show that φ is an antiisomorhism between $\mathcal{F}(C)$ and $\mathcal{F}(C^P)$ in several steps.

(1) $\varphi : \mathcal{F}(C) \to \mathcal{F}(C^P)$: Let F be a face of C. Then every $c \in F$ defines a valid inequality $c^T x \leq 0$ for C^P. Hence $C^P \cap c^\perp$ is a face of C^P. Therefore,

$$\varphi(F) = C^P \cap F^\perp = \bigcap \{C^P \cap c^\perp \mid c \in F\}$$

is an intersection of faces of C^P, and hence is itself a face of C^P.

(2) $F \leq F'$ implies $\varphi(F) \geq \varphi(F')$: This is obvious from the definition of φ.

(3) φ is injective: Define $\bar{\varphi} : \mathcal{F}(C^P) \to \mathcal{F}(C)$ by

$$\bar{F} \to C \cap \bar{F}^{\perp}.$$

Note that it follows from 1) that $\bar{\varphi}$ actually maps $\mathcal{F}(C^P)$ to $\mathcal{F}(C)$. Now 3) will follow, once we have shown that $\bar{\varphi} \circ \varphi = \text{id}$. To see this, let $F \in \mathcal{F}(C)$. Then, by Lemma 7.9 and Corollary 7.11, we get

$$
\begin{aligned}
(C^P \cap F^{\perp})^{\perp} &= (-C^P \cap F^{\perp})^P \cap (C^P \cap F^{\perp})^P \\
&= ((C^P \cup -C^P) \cap F^{\perp})^P \\
&= (cone(C^P \cup -C^P) \cap F^{\perp})^P \\
&= (lin(C^P) \cap F^{\perp})^P = ((lineal\, C)^{\perp} \cap F^{\perp})^P \\
&= (F^{\perp})^P = F^{\perp\perp} = lin\, F
\end{aligned}
$$

Hence

$$\bar{\varphi}\varphi(F) \quad = \quad C \cap (C^P \cap F^{\perp})^{\perp} = C \cap lin\, F = F.$$

(4) φ is surjective: We have just seen that $\bar{\varphi} \circ \varphi = \text{id}$. By symmetry (i.e. replacing C by C^P), we conclude that $\varphi \circ \bar{\varphi} = \text{id}$, too. Hence φ is surjective.

This completes the proof of the theorem.

$\square$

Lemma 7.42 Let $C \subseteq \mathbb{K}^n$ be a polyhedral cone, and let C^P its polar. Then the antiisomorphism

$$\varphi := F \to \bar{F} = C^P \cap F^{\perp}$$

is such that $\dim F + \dim \bar{F} = n$ for every face F of C.

Proof. It suffices to show that $lin\, \bar{F} = F^{\perp}$ for every face F of C. To see this, first note that, as in step (3) in the proof of Theorem 7.41,

$$(C^P \cap F^{\perp})^{\perp} = lin\, F.$$

Hence

$$lin\, \bar{F} = lin(C^P \cap F^{\perp}) = (C^P \cap F^{\perp})^{\perp\perp} = (lin\, F)^{\perp} = F^{\perp}.$$

$\square$

Corollary 7.43 *If $C \subseteq \mathbb{K}^n$ is a polyhedral cone, and F and F' are two faces of C such that F is a facet of F', i.e. F' covers F in the face lattice of C, then* $\dim F = \dim F' - 1$.

Proof. W.l.o.g., suppose that $F' = C$, i.e. that F is a facet of C. Since $\varphi := F \to \bar{F}$ is an antiisomorphism, this implies that $\bar{C}$, the minimal face of C^P, is a facet of $\bar{F}$. Thus, by Lemma 7.42, we are left to show the following: If $C \subseteq \mathbb{K}^n$ is a polyhedral cone with minimal face F and a face F' covering F, then $\dim F = \dim F' - 1$. This is obvious in case C is pointed, for in this case $F = \{0\}$, and F' must be an extreme ray $\mathbb{K}_+ e$ of C. The general case, however, is easily derived from the special case of pointed cones. In fact, let $C \subseteq \mathbb{K}^n$ be an arbitrary polyhedral cone and consider the associated pointed cone $C_0 = C \cap \mathit{lineal}\, C^\perp$. Then F is a face of C if and only if $F = F_0 + \mathit{lineal}\, C$ for some face F_0 of C_0, implying that $\dim F' - \dim F = \dim F_0' - \dim F_0$ for any two faces F, F' of C. This proves the claim.

□

Theorem 7.41 states that the face lattices $\mathcal{F}(C)$ and $\mathcal{F}(C^P)$ of a polar pair of polyhedral cones are, in a sense, the same thing. Now consider the oriented matroid $\mathcal{O} = \sigma(\mathit{im}\, B)$ and the set $T = \{Y \in \mathcal{O} \mid Y \leq 0\}$. We know already that T and $\mathcal{F}(C)$ are isomorphic. Hence T and $\mathcal{F}(C^P)$ are antiisomorphic. Hence both $\mathcal{F}(C)$ and $\mathcal{F}(C^P)$ can be considered as "geometrical realizations" of the abstract, purely "combinatorial" lattice $(T, \preceq)$. When we study such "tope lattices" in a more abstract setting in Chapter 8, it will be helpful to associate to T intuitively either one of the two possible "realizations" C and C^P (although, in general, i.e. when $\mathcal{O}$ does not arise from a subspace of $\mathbb{K}^n$, such realizations may not exist). It does not matter at all, whether we think of C or C^P when working with T as above, however, it will be less confusing to fix one of them arbitrarily. We decided to fix C. Thus given $\mathcal{O} = \sigma(\mathit{im}\, B)$ and T as above, we will always interprete the order $\preceq$ on T as the inclusion order on the faces of $C = P(B, 0)$.

7.8 General Polyhedra

It is not difficult to generalize the results of the preceding sections for polyhedral cones to general polyhedra. It is not very interesting either, but we would like to do so, just for the sake of completeness. Of course the basic tool to be applied is (de-) homogenization.

Let us start with defining extreme points and faces of general polyhedra.

Definition 7.44 *Let $P \subseteq \mathbb{K}^n$ be a polyhedron, and let $C \subseteq \mathbb{K}^{n+1}$ denote its homogenization. Then $x \in P$ is called* **extreme** *if the corresponding vector $\binom{x}{1} \in C$ is extreme. Extreme points of polyhedra are also called* **vertices**.

It is straightforward from the definition that $x \in P$ is a vertex if and only if x is not a proper convex combination of $x', x'' \in P$.

Example 7.45

1. *Let $P = \text{lineal}\, P$ be a subspace of $\mathbb{K}^n$. Then P has no vertices.*
2. *Let $P \subseteq \mathbb{K}$ be defined by $P := \{x \in \mathbb{K} \mid 0 \leq x \leq 1\}$ then $x_0 = 0$ and $x_1 = 1$ are all vertices of P.*
3. *Let $P \subseteq \mathbb{K}^2$ be defined by $P = \{x \in \mathbb{K}^2 \mid x_1 \geq 0\}$. Again, P has no vertices.*
4. *Let $P \subseteq \mathbb{K}^2$ be defined by $P := \mathbb{K}^2_+$. Then $x = 0$ is the only vertex of P.*

Theorem 7.46 (MINKOWSKI) *Let $P \subseteq \mathbb{K}^n$ be a nonempty polyhedron. Then P can be represented as*

$$P = \text{conv}\, V + \text{cone}\, E$$

where V and E are finite subsets of $\mathbb{K}^n$. In fact, V can be chosen to be the set of vertices of P and E can be chosen to be any set generating $\text{rec}\, P$, i.e. such that $\text{cone}\, E = \text{rec}\, P$.

Proof. Let $P \subseteq \mathbf{K}^n$ be given. Its homogenization is

$$\tilde{P} = \left\{\lambda \begin{pmatrix} x \\ 1 \end{pmatrix} \;\middle|\; x \in P\right\} \cup \left\{\begin{pmatrix} x \\ 0 \end{pmatrix} \;\middle|\; x \in \text{rec}\, P\right\}.$$

By MINKOWSKI's Theorem (cf. Theorem 7.6), $\tilde{P}$ is the conic hull of some finite set $\tilde{E} \subseteq \mathbf{K}^{n+1}$. By scaling the elements of $\tilde{E}$ with appropriate positive factors, we may assume that every element of $\tilde{E}$ has last coordinate equal to 1 or 0. Hence $\tilde{E}$ can be written as

$$\tilde{E} = \begin{pmatrix} V \\ 1 \end{pmatrix} \cup \begin{pmatrix} E \\ 0 \end{pmatrix}$$

for some finite sets $V \subseteq \mathbf{K}^n$ and $E \subseteq \mathbf{K}^n$.

Now it is easy to see that the dehomoganization of $\tilde{P} = \text{cone}\, \tilde{E}$ is equal to

$$P = \text{conv}\, V + \text{cone}\, E.$$

The second claim follows from the decomposition

$$\tilde{P} = cone\,\tilde{E} = cone\binom{V}{1} + cone\binom{E}{0}.$$

In fact,

$$rec\,P = \left\{ x \;\middle|\; \binom{x}{0} \in \tilde{P} \right\} = cone\,E,$$

showing that E can be replaced by any set generating $rec\,P$. Finally, $cone\binom{V}{1}$ is generated by its extreme rays which are in 1-1 correspondence to vertices of P.

□

Theorem 7.47 (WEYL) *Let V and E denote finite subsets of $\mathbb{K}^n$. Then*

$$P = conv\,V + cone\,E$$

is a polyhedron.

Proof. The set

$$\tilde{P} = cone\binom{V}{1} + cone\binom{E}{0}$$

is a polyhedral cone whose dehomogenization equals P.

□

Next, let us introduce faces of general polyhedra in the same way as we defined this concept for polyhedral cones. Note, however, that this time we define the dimension of a face F via its affine hull rather than its linear hull. This is compatible with the corresponding definition for faces of polyhedral cones, since $aff\,F = lin\,F$ whenever F is a polyhedral cone. Furthermore, note that if $P \subseteq \mathbb{K}^n$ is a polyhedron, then, according to the definition below, the empty set is a face of P. This contrasts the corresponding definition for polyhedral cones, because, by definition, a polyhedral cone $C \subseteq \mathbb{K}^n$, and hence any face of C, is nonempty. Thus, whether $\emptyset$ is a face of C, depends on whether we regard C as a polyhedron or as a polyhedral cone. This may look somewhat irritating at the first glance. However, no severe misunderstandings are possible and therefore we decided to accept this slight incompatibility. On the positive side, our definitions yield that for a polytope $P \subseteq \mathbb{K}^n$, the "face lattice" of P is isomorphic to the face lattice of its homogenization $\tilde{P} \subseteq \mathbb{K}^{n+1}$ (cf. Proposition 7.56 for the case of unbounded polyhedra).

Definition 7.48 *Let $P \subseteq \mathbb{K}^n$ be polyhedron. Then a polyhedron $F \subseteq P$ is called a* **face** *of P, provided the following holds: If $x \in F$ is a* *proper convex combination of $x', x'' \in P$ then both $x', x'' \in F$. A face F of P is called* **proper** *if $F \subset P$. Maximal proper faces are called* **facets.** *Any face consisting of a single point x is called a* **vertex** *of P. The* **dimension** *of a face F is defined to be $\dim F := \dim(\mathrm{aff} F)$.*

Again, faces can be characterized as intersections of P with "supporting hyperplanes":

Definition 7.49 *Let $P \subseteq \mathbb{K}^n$ be a polyhedron, and let $c^T x \leq c_0$ be a valid inequality. Then $H = \{x \mid cx = c_0\}$ is called a* **supporting hyperplane** *for P. Note that $c = 0, c_0 = 0$ is not excluded.*

In case P is given by a system of linear inequalities $Ax \leq b$, the inequalities $A_{i.}x \leq b_i$ provide a natural set of valid inequalities. Proceeding in analogy to the conic case, we introduce equality sets as follows.

Definition 7.50 *Let $P = P(A, b)$ for some $A \in \mathbb{K}^{m\times n}, b \in \mathbb{K}^m$, and let $x \in P$. Then*

$$Q(x) := \{i \mid A_{i.}x = b_i\}$$

is called the **equality set** *of x. More generally, for $S \subseteq P$ let*

$$Q(S) := \bigcap_{x \in S} Q(S)$$

denote the **equality set** *of S. If $F \subseteq P$ is a polyhedron, then $x \in F$ is called an* **interior point** *of F, if $Q(x) = Q(F)$.*

This obviously generalizes the corresponding definition for polyhedral cones, and we therefore refer to earlier remarks justifying the terminology. In fact, the characterization of faces by means of supporting hyperplanes can be derived for general polyhedra in the very same way as for polyhedral cones. Therefore, it may be left to the reader to prove the following analogue of Theorem 7.27.

Theorem 7.51 *Let $P = P(A, b)$ be a polyhedron in $\mathbb{K}^n$. Then $F \subseteq P$ is a face of P if and only if*

$$F = P \cap \{x \in \mathbb{K}^n \mid A_{i.}x = b_i \ \forall i \in Q(F)\}.$$

Every face F of P can be written as $F = P \cap H$ where H is some supporting hyperplane of P.

□

Example 7.52 *Consider the map $x : I\!K_+ \to I\!K^n$, defined by $x(t) = (t, t^2, \ldots, t^n)$. Then*

$$M_n := x(I\!K_+) = \{x(t) \mid t \in I\!K_+\}$$

is called the **moment curve** *in $I\!K^n$. If $r > n$, then*

$$C(r,n) := conv\{x(i) \mid i = 1, \ldots, r\}$$

is called the **cyclic polytope of dimension** n **with** r **vertices.** *It is not difficult to see that each of the $x(i)$ is a vertex. In fact, one can show that for $k \leq \frac{n}{2}$, every subset $\{i_1, \ldots, i_k\} \subseteq \{1, \ldots, r\}$ yields a face*

$$F = conv\{x(i_1), \ldots, x(i_k)\}$$

of $C(r,n)$.

To see this, consider the polynomial

$$p(t) = \prod_{j=1}^{k} (t - i_j)^2 = c_0 + c_1 t + \ldots + c_{2k} t^{2k}$$

where the coefficients c_m depend only on the i_j's. Let

$$c := (c_1, \ldots, c_{2k}, 0, \ldots, 0) \in I\!K^n$$

and

$$H := \{x \in I\!K^n \mid c^T x = -c_0\}.$$

Then obviously,

$$c_0 + c^T x(i_j) = c_0 + c_1 i_j + c_2 i_j^2 + \ldots + c_{2k} i_j^{2k} = p(i_j) = 0.$$

Thus $x(i_j) \in H$ for every $j = 1, \ldots, k$.

On the other hand, for every $i \notin \{i_1, \ldots, i_k\}$ we get

$$0 < p(i) = c_0 + c^T x(i).$$

Thus H is a supporting hyperplane defining the face $F = C(r,n) \cap H$.

It is easy to see that $F = conv\{x(i_1), \ldots, x(i_k)\}$ is a $k-1$ dimensional face of $C(r,n)$. Hence the above argument shows that the number of k-dimensional faces, $k \leq \frac{n}{2} - 1$, is given by

$$f_k = \binom{r}{k+1}$$

which is obviously as large as possible.

Having a large number of faces is the most important property of cyclic polytopes. In fact, one can prove the socalled "Upper Bound Theorem", stating that cyclic polytopes have the maximum possible number of k-faces for every k, among all polytopes of a given dimension n with a given number $r > n$ of vertices (cf. the references in Section 7.9).

Definition 7.53 *Let $P \subseteq \mathbb{K}^n$ be a polyhedron, and let $\mathcal{F} = \mathcal{F}(P)$ denote the set of faces of P. For $F, F' \in \mathcal{F}$, let $F \leq F'$ if F is a face of F' (or, equivalently, $F \subseteq F'$). Then $(\mathcal{F}, \leq)$ is a lattice, called the* **face lattice** *of P.*

We warn the reader that the face lattice of a polyhedron is in general not isomorphic to the face lattice of its homogenization. This can be seen easily by considering, e.g. $P = \mathbf{K}_+$ and $C = \mathbf{K}_+^2$. Before we can investigate the relationship between face lattices of polyhedra and their homogenizations, a few remarks on homogenization and dehomogenization are in order. In the following, if $P \subseteq \mathbf{K}^n$ is a polyhedron, then $\tilde{P} \subseteq \mathbf{K}^{n+1}$ denotes its homogenization.

Lemma 7.54 *Homogenization is injective and preserves inclusion and nonempty intersections.*

Proof. Shortening the notation in an obvious way, let us write

$$\tilde{P} = cone\begin{pmatrix} P \\ 1 \end{pmatrix} + \begin{pmatrix} rec\, P \\ 0 \end{pmatrix}$$

for the homogenization of a polyhedron $P \subseteq \mathbf{K}^n$. It is obvious that homogenization is injective and preserves inclusion. Furthermore, if $P \cap Q \neq \emptyset$, then

$$\begin{aligned} \tilde{P} \cap \tilde{Q} &= \left[cone\begin{pmatrix} P \\ 1 \end{pmatrix} + \begin{pmatrix} rec\, P \\ 0 \end{pmatrix}\right] \cap \left[cone\begin{pmatrix} Q \\ 1 \end{pmatrix} + \begin{pmatrix} rec\, Q \\ 0 \end{pmatrix}\right] \\ &= cone\begin{pmatrix} P \\ 1 \end{pmatrix} \cap cone\begin{pmatrix} Q \\ 1 \end{pmatrix} + \begin{pmatrix} rec\, P \\ 0 \end{pmatrix} \cap \begin{pmatrix} rec\, Q \\ 0 \end{pmatrix} \\ &= cone\begin{pmatrix} P \cap Q \\ 1 \end{pmatrix} + \begin{pmatrix} rec(P \cap Q) \\ 0 \end{pmatrix} = (P \cap Q)^{\sim}. \end{aligned}$$

□

Lemma 7.55 *Let $P \subseteq \mathbb{K}^n$ be a polyhedron. Then the following holds:*

(i) *If $F \leq P$ is nonempty then $\tilde{F} \leq \tilde{P}$.*

(ii) *If* $F \leq P$ *then* $\begin{pmatrix} rec\,F \\ 0 \end{pmatrix} \leq \tilde{P}$.

(iii) *Every face of* $\tilde{P}$ *can be obtained as in (i) or (ii).*

Proof.

ad (ii): Let $F \leq P$. Then $rec\,F \leq rec\,P$. In fact, if H is any supporting hyperplane of P such that $F = P \cap H$, then $rec\,H$ is a supporting hyperplane of $rec\,P$ such that $rec\,F = (rec\,P) \cap (rec\,H)$. Hence $rec\,F$ is a face of $rec\,P$.

This further implies that $\binom{rec\,F}{0}$ is a face of $\binom{rec\,P}{0}$. But $\binom{rec\,P}{0}$ obviously is a face of $\tilde{P}$ (with supporting hyperplane $x_{n+1} = 0$). By transitivity of the relation $\leq$, this implies that $\binom{rec\,F}{0} \leq \tilde{P}$.

ad (i): Let $F \leq P$. Let $c^T x \leq c_0$ be valid for P such that

$$F = P \cap H, \text{ where } H = \{x \in \mathbf{K}^n \mid c^T x = c_0\}.$$

The homogenization $\tilde{H}$ of H is given by

$$\tilde{H} = cone \begin{pmatrix} H \\ 1 \end{pmatrix} + \begin{pmatrix} rec\,H \\ 0 \end{pmatrix} = \{\tilde{x} \in \mathbf{K}^{n+1} \mid (c, -c_0)^T \tilde{x} = 0\}.$$

From this representation, one can easily see that $\tilde{H}$ is a supporting hyperplane of $\tilde{P}$. Hence $\tilde{P} \cap \tilde{H}$ is a face of $\tilde{P}$. Since homogenization preserves nonempty intersection by Lemma 7.54 above, we get

$$\tilde{F} = (P \cap H)^{\sim} = \tilde{P} \cap \tilde{H}.$$

Hence $\tilde{F}$ is a face of $\tilde{P}$.

ad (iii): Let $\tilde{F} \leq \tilde{P}$. We consider two cases:

a) $\tilde{F} \subseteq \mathbf{K}^n \times \{0\}$, i.e. $\tilde{F} = \binom{R}{0}$ for some set $R \subseteq \mathbf{K}^n$. Then $\tilde{F} \subseteq \tilde{P} \cap (\mathbf{K}^n \times \{0\}) = \binom{rec\,P}{0}$. Hence $\tilde{F} = \binom{R}{0}$ is a face of $\binom{rec\,P}{0}$, implying that R is a face of $rec\,P$. We have to show that this further implies $R = rec\,F$ for some face $F \leq P$. This can be seen as follows. Let

$$P = conv\,V + rec\,P$$

for some nonempty finite set $V \subseteq \mathbf{K}^n$. Let $c^T x \leq 0$ be valid for $rec\,P$ such that

$$R = H \cap rec\,P, \text{ where } H = \{x \mid c^T x = 0\}.$$

Let $c_0 := \max\{c^T v \mid v \in V\}$. Then $c^T x \leq c_0$ is valid for P and $F = P \cap \{x \mid c^T x = c_0\}$ is a face of P satisfying:

$$rec\, F = (rec\, P) \cap rec\{x \mid c^T x = c_0\} = (rec\, P) \cap H = R.$$

b) $\tilde{F} \not\subseteq \mathbf{K}^n \times \{0\}$. In this case we will see that the dehomogenization F of $\tilde{F}$ is a face of P. Since homogenizing F leads back to $\tilde{F}$, this proves that $\tilde{F}$ is indeed a homogenized face of P. $\tilde{F}$ may be written as

$$\tilde{F} = cone\begin{pmatrix} F \\ 1 \end{pmatrix} + \begin{pmatrix} rec\, F \\ 0 \end{pmatrix},$$

while $\tilde{P}$ equals

$$\tilde{P} = cone\begin{pmatrix} P \\ 1 \end{pmatrix} + \begin{pmatrix} rec\, P \\ 0 \end{pmatrix}.$$

Now $\tilde{F} \leq \tilde{P}$ implies that $\tilde{F} \cap (\mathbf{K}^n \times \{1\}) \leq \tilde{P} \cap (\mathbf{K}^n \times \{1\})$. Hence $\binom{F}{1} \leq \binom{P}{1}$, implying that $F \leq P$.

□

The preceding result gives a precise description of the relationship between a polyhedron P and its homogenization $\tilde{P}$: The faces of $\tilde{P}$ which are contained in $\mathbf{K}^n \times \{0\}$ are precisely the faces of $rec\, P$. The remaining ones are homogenized faces of P. An easy consequence is the following:

Proposition 7.56 *Let $P \subseteq \mathbb{K}^n$ be a polyhedron, and let $\tilde{P}$ be its homogenization. Then $\mathcal{F}(P)$ is isomorphic to*

$$\tilde{\mathcal{F}} = \{\tilde{F} \in \mathcal{F}(\tilde{P}) \mid \tilde{F} = 0 \text{ or } \tilde{F} \not\subseteq \mathbb{K}^n \times \{0\}\}.$$

Hence, in particular, if P is bounded then $\mathcal{F}(P) \cong \mathcal{F}(\tilde{P})$.

Proof. We know already that homogenization gives rise to a bijective map $F \to \tilde{F}$ between $\mathcal{F}(P)$ and $\tilde{\mathcal{F}}$ as above ($\tilde{F} = 0$ is the image of the empty face of P). Furthermore, homogenization preserves inclusion and nonempty intersection, hence we have

$$F' \leq F'' \Leftrightarrow \tilde{F}' \leq \tilde{F}'' \text{ and } (F' \wedge F'')^\sim = \tilde{F}' \wedge \tilde{F}''.$$

(Note that if $F' \wedge F'' = \emptyset$ then $\tilde{F}' \wedge \tilde{F}''$ is indeed 0 in $\tilde{\mathcal{F}}$.) Since $F' \vee F''$ is the intersection of all faces containing both F' and F'', this also implies that $(F' \vee F'')^\sim = \tilde{F}' \vee \tilde{F}''$.

□

Note that Proposition 7.56 states that the face lattice of a polyhedron P is essentially (i.e. if we forget about the zero face of $\tilde{P}$) the same as the face lattice of $\tilde{P}$, with all faces $\tilde{F} \subseteq \mathbb{K}^n \times \{0\}$ removed. Now $\mathbb{K}^n \times \{0\}$ is a supporting hyperplane of $\tilde{P}$ and hence

$$\tilde{F}_0 = \tilde{P} \cap (\mathbb{K}^n \times \{0\})$$

is a face of $\tilde{P}$. Thus — "essentially" —, the face lattice of P equals the face lattice of $\tilde{P}$ minus all faces of $\tilde{F}_0$, i.e. $\mathcal{F}(P) \approx \mathcal{F}(\tilde{P}) \setminus \mathcal{F}(\tilde{F}_0)$. Of course, there is nothing special about $\mathbb{K}^n \times \{0\}$ and $\tilde{F}_0$. Given any polyhedral cone $\tilde{P} \subseteq \mathbb{K}^{n+1}$ and any face $\tilde{F}_0$ of $\tilde{P}$, we may assume w.l.o.g that

$$\tilde{F}_0 = \tilde{P} \cap (\mathbb{K}^n \times \{0\}).$$

(Otherwise, we may apply a linear transformation, mapping $\mathbb{K}^n \times \{0\}$ to a supporting hyperplane H of $\tilde{F}_0$.) Hence, removing a face (together with its subfaces) from a polyhedral cone essentially gives a face lattice of a polyhedron. This can be seen as an analogue of the corresponding result from linear algebra: Removing a hyperplane from a projective space yields an affine space.

Finally, let us see what "removing a face" means in the oriented matroid terminology. Let $P = P(A, b) \subseteq \mathbb{K}^n$. Its homogenization C is given by $C = P(B, 0)$ where

$$B = \begin{pmatrix} A & -b \\ 0 & -1 \end{pmatrix}.$$

Let $\mathcal{O} = \sigma(im\, B)$ and $T = \{Y \in \mathcal{O} \mid Y \leq 0\}$. We know (cf. Section 7.5) that the face lattice of C is isomorphic to $(T, \preceq)$. The isomorphism is given by the canonical map

$$F \to \sigma \circ B(int\, F).$$

Removing the face $C \cap (\mathbb{K}^n \times \{0\})$ from C (together with all its subfaces) therefore amounts to removing all $Y = \sigma(Bx)$ with $x \in \mathbb{K}^n \times \{0\}$. What is left then is the "dehomogenized" or "affine" version

$$T_{aff} = \{Y \in \mathcal{O} \mid Y \leq 0 \text{ and } Y_{m+1} = -\}$$

of T. It follows from our considerations above that $T_{aff} \cup \{0\}$ is isomorphic to $\mathcal{F}(P)$.

More generally, let $C = P(B, 0)$ be an arbitrary polyhedral cone with $B \in \mathbb{K}^{m \times n}$. Then $\mathcal{O} = \sigma(im\, B)$ is an oriented matroid on $\{1, \ldots, m\}$. Removing the face determined by the i-th inequality results in

$$T'_{aff} = \{Y \in \mathcal{O} \mid Y \leq 0 \text{ and } Y_i = -\}.$$

Of course, in general, C has much more faces than those determined by a single inequality $B_{i.}x \leq 0$. Thus, in general, if F is a face of C then removing F results in

$$T_{aff}^{Q(F)} = \{Y \in \mathcal{O} \mid Y \leq 0 \text{ and } Y_i = - \text{ for some } i \in Q(F)\}$$

where $Q(F)$, as in Section 7.3, denotes the equality set of F.

Summarizing, we can say that the relationship between face lattices of polyhedral cones and those of general polyhedra is, though not completely trivial, at least quite well understood. This may be taken as a justification for restricting our attention to face lattices of polyhedral cones and face lattices of oriented matroids. First, however, let us finish this section with a quick look at polarity between general polyhedra, just for the sake of completeness.

There is a natural way to define the polar of a polyhedron $P \subseteq \mathbb{K}^n$. First, let $C \subseteq \mathbb{K}^n \times \mathbb{K}_+$ denote its homogenization. Next construct C^P, the polar of C and define the polar P^* of P to be the negative dehomogenization (cf. Section 4.2) of C^P.

Definition 7.57 *Let $P \subseteq \mathbb{K}^n$ be a polyhedron. Then the polar of P, denoted by $P^* \subseteq \mathbb{K}^n$, is defined to be the negative dehomogenization of the polar of the homogenization of P.*

What a bombastic definition! One is almost tempted to regret that it all boils down to the following

Proposition 7.58 *Let $P \subseteq \mathbb{K}^n$ be a polyhedron. Then*

$$P^* = \{y \in \mathbb{K}^n \mid x^T y \leq 1 \; \forall x \in P\}.$$

Proof. Let C denote the homogenization of P.

Then $y \in P^* \Leftrightarrow \binom{y}{-1} \in C^P \Leftrightarrow (y, -1) \cdot \tilde{x} \leq 0 \quad \forall \tilde{x} \in C$. Now $C = cone\binom{P}{1} \cup \binom{rec\,P}{0}$, hence

$$\begin{aligned} y \in P^* \;\Leftrightarrow\; & y^T x \leq 1 \quad \forall x \in P \quad \text{and} \quad y^T u \leq 0 \quad \forall u \in rec\,P \\ \Leftrightarrow\; & y^T x \leq 1 \quad \forall x \in P, \end{aligned}$$

since obviously $yx \leq 1 \quad \forall x \in P$ implies $y^T u \leq 0 \quad \forall u \in rec\,P$.

□

In particular, if P happens to be a polyhedral cone, then P^* equals the polar cone as defined earlier, since $y^T x \leq 1 \quad \forall x \in P$ implies $y^T x \leq 0 \quad \forall x \in P$, in case P is a cone.

The concept of polarity for general polyhedra is, however, somewhat less straightforward, compared with the corresponding homogenized notion. In particular, the shape of P^* may depend rather drastically on the position of P in $\mathbb{K}^n$, as the following example shows.

Example 7.59

(i) *Let* $P = [-1,1]^2 \subseteq \mathbb{K}^2$. *Then* P^* *is given by*

$$P^* = conv\left\{\begin{pmatrix}1\\0\end{pmatrix}, \begin{pmatrix}-1\\0\end{pmatrix}, \begin{pmatrix}0\\1\end{pmatrix}, \begin{pmatrix}0\\-1\end{pmatrix}\right\}$$

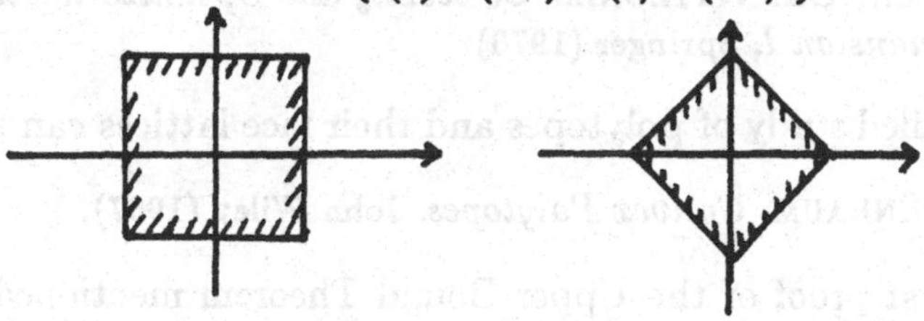

(ii) *Let* $P = [0,2]^2 \subseteq \mathbb{K}^2$. *Then* P^* *is given by*

$$P^* = \left\{\begin{pmatrix}x\\y\end{pmatrix} \in \mathbb{K}^2 \mid \quad x + y \leq \frac{1}{2}, x \leq \frac{1}{2}, y \leq \frac{1}{2}\right\}$$

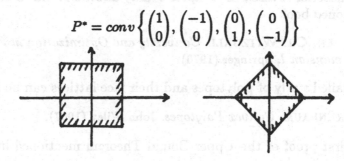

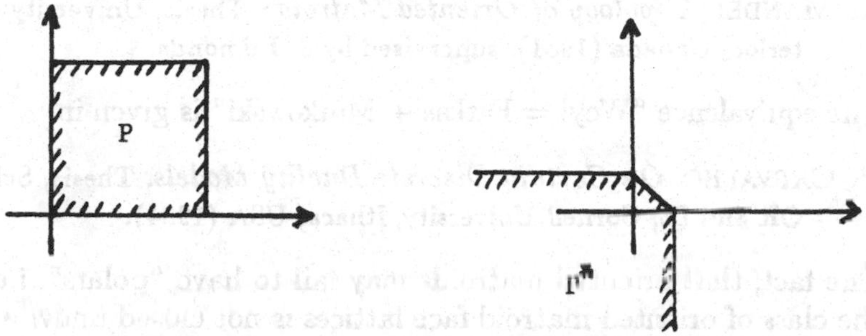

The explanation for this phenomenon is, roughly, that the negative dehomogenization of C^P works well only if $C^P \subseteq \{x \in \mathbb{K}^{n+1} \mid x_{n+1} < 0\} \cup \{0\}$. This amounts to say that the $(n+1)$-th unit vector e_{n+1} is in the interior of C, i.e. $0 \in int\, P$. Of course, $0 \in int\, P$ implies $0 \in int\, P^*$. Hence one may restrict the polar operation to the class of polyhedra whose interior contains the origin. Within this restricted class (which, of course, means no loss of generality as far as face lattices are concerned), the polar operator $*$ behaves nicely. In fact, one can show that $P^{**} = P$ and that the face lattices of P and P^* are antiisomorphic. We won't do so, however, since, after all, one doesn't get anything conceptually different from the homogenized case.

Thus, let us stop our investigation of general polyhedra here and turn back into the homogenous world of cones, linear spaces, and oriented matroids!

7.9 Further Reading

The material covered in Chapter 7 may also be found in the earlier mentioned book

J. STOER, CH. WITZGALL *Convexity and Optimization in Finite Dimension I*, Springer (1970).

A detailed study of polytopes and their face lattices can be found in

B. GRÜNBAUM *Convex Polytopes*, John Wiley (1967).

The first proof of the Upper Bound Theorem mentioned in Section 7.8 was given by

P. MCMULLEN *The Maximum Number of Faces of a Convex Polytope*, Mathematica **17** (1970), pp. 179–184.

A generalization of this result, stating that the Upper Bound Theorem also holds for face lattices of oriented matroids, is proved in

A. MANDEL *Topology of Oriented Matroids*, Thesis, University of Waterloo, Canada (1981), supervised by J. Edmonds.

The equivalence "Weyl = Farkas + Minkowski" is given in

P. CARVALHO *On Certain Discrete Duality Models*, Thesis, School of OR and IE, Cornell University, Ithaca, USA (1984).

The fact, that oriented matroids may fail to have "polars", i.e. that the class of oriented matroid face lattices is not closed under antiisomorphism, is proved in

B. MUNSON *Face Lattices of Oriented Matroids*, Thesis, Cornell University, Ithaca, USA (1981),

cf. also

L.J. BILLERA, B. MUNSON *Polarity and Inner Products in Oriented Matroids*, Techn. Report, Cornell University, Ithaca, USA (1981).

Chapter 8

The Poset $(\mathcal{O}, \preceq)$

In Chapter 6 we have analyzed the structure of an oriented matroid $\mathcal{O}$, considered as a system of sign vectors. Here we will investigate its structure as a poset. These two points of view are strongly related, of course, though the relationship is not as clear as one might expect at the first glance. For example, if a set of sign vectors is given and we are to decide whether this is an OM, then we may simply check the axioms in order to find out the answer. On the other hand, if we are given a poset $(\mathcal{P}, \preceq)$, there is no obvious way to decide wether it is an OM-poset or not, except by "brute force", i.e. by trying all possible OMs up to a certain size and each time comparing their posets to the given one. [21] and [48] provide a more clever way of computing a set $\mathcal{O}$ of sign vectors such that $(\mathcal{O}, \preceq)$ equals the given poset $(\mathcal{P}, \preceq)$. But nonetheless, the final check whether or not $(\mathcal{O}, \preceq)$ is an OM poset has to be done by checking the sign vector axioms. The reason is that, so far, there is no neat characterization of OMs in terms of their posets.

This chapter presents what is currently known about structural properties of OM posets. Section 8.1 provides some "without loss" simplifications. Section 8.2 developes some basic results, e.g. the existence of a dimension function and a proof that OM face lattices are relatively complemented. Sections 8.3 and 8.4 deal with constructability of OM-posets. This is probably one of the most interesting properties of OM posets, which roughly states that an OM poset can be obtained by successively pasting together small parts of it. This result is due to J. Edmonds and A. Mandel [126]. Constructibility will turn out to be an important tool for constructing topological realizations of OMs in Chapter 9.

8.1 Simplifications

This section is to provide basic tools for simplifying the exposition. Let us introduce some notation first. To motivate the definition, recall the concept of the canonical map from Section 7.5. Consider a linear OM $\mathcal{O} = \sigma(im\, B)$ as in Section 7.5. Then the canonical map $\sigma \circ B : \mathbb{K}^n \to \mathcal{O}$ provides a 1-1 correspondence between cones C_Y definable from $Bx \le 0$ and their corresponding sets $[Y] = \{X \in \mathcal{O} \mid X \preceq Y\}$. This correspondence is given by $[Y] = \sigma \circ B(C_Y)$. Furthermore, the correspondence is such that $X \preceq Y$, i.e. $[X] \subseteq [Y]$ iff $C_X \le C_Y$, i.e. C_X is a face of C_Y. Recall from Section 7.5 that the C_Y's were also called "cells".

Definition 8.1 *Let $\mathcal{O}$ be an OM on some set E. If $Y \in \mathcal{O}$ then $[Y] := \{X \in \mathcal{O} \mid X \preceq Y\}$ is called a* **cell**. *If $[X] \subseteq [Y]$ then $[X]$ is called a* **face** *of $[Y]$. In case $[X] \neq [Y]$, $[X]$ is called a* **proper face**. *A maximal proper face is called a* **facet**. *If $[Y]$ is a cell and $[X]$ is a facet of $[Y]$ such that $X_e = 0$ for some $e \in supp\, Y$, then $[X]$ is the unique facet of $[Y]$ satisfying $X_e = 0$. (In fact, X is the conformal sum of all $Z \preceq Y$ having $Z_e = 0$.) This $[X]$ is called the facet of $[Y]$* **determined** *by e and e is called a* **facet equator** *of $[X]$. (Note that e is not necessarily uniquely determined by $[X]$.) If we want to stress that we consider $[X]$ as a poset, then we will also talk about the* **face lattice** *of $[X]$.*

If $X \in \mathcal{O}$ is elementary, then $[X]$ is called a **vertex**. *(Thus vertices are minimal cells $\neq [0]$.) If Y covers an elementary X in $\mathcal{O}$, then $[Y]$ is called an* **edge**. *A maximal cell $[T]$ is also called a* **tope**. *Face lattices of topes are also called* **tope lattices**, *for short.*

Consider Figure 8.1 below which is to be understood in the same way as Figure 7.5 of Section 7.5. In the configuration indicated, $[X]$ and $[X']$ are vertices. $[Y]$ and $[Y']$ are edges, and $[T]$ and $[T']$ are topes. $[Y]$ is a facet of $[T]$ with facet equators e and $\bar{e}$. $[Y']$ is also a facet of $[T]$ but has only a single facet equator f. $[X']$ is a facet of $[Y']$ with facet equators g and h.

Remark 8.2 Note that if $\mathcal{O} = \sigma(im\, B)$ and $[X]$ is a vertex, then the dimension of $C_X = (\sigma \circ B)^{-1}[X]$ is not necessarily equal to 1. In fact, this is only true if $ker\, B = 0$. However, since we are interested only in $\mathcal{O} = \sigma(im\, B)$ rather than the matrix B itself, we may assume w.l.o.g. that $ker\, B = 0$. (Otherwise remove some columns of B until the remaining matrix has full column rank.)

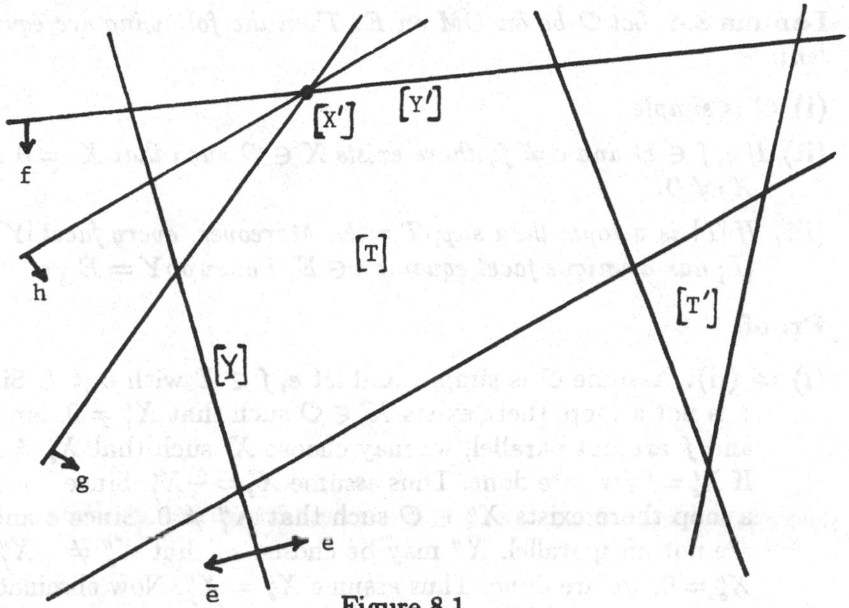

Figure 8.1

If $\mathcal{O} = \sigma(im\, B)$ with $ker\, B = 0$, then the canonical map relates vertices and edges to 1- and 2-dimensional cones, resp. More generally, we may define $\dim : \mathcal{O} \to \mathbb{Z}$ by $\dim X = \dim C_X - 1$. Thus $\dim 0 = -1$, and vertices and edges have dimension 0 and 1, resp. More generally, if $[X]$ is a facet of $[Y]$, i.e. Y covers X in $\mathcal{O}$, then $\dim Y = \dim X + 1$, as follows from the corresponding relation between polyhedral cones (cf. Corollary 7.43). As we will see in Section 8.2, a similar concept of "dimension" can be introduced for general OM posets, i.e. also in case $\mathcal{O}$ is not given by $\mathcal{O} = \sigma(im\, B)$ for some matrix B.

Definition 8.3 *Let $\mathcal{O}$ be an OM on E. Then $e \in E$ is called a* **loop** *if $X_e = 0$ for every $X \in \mathcal{O}$. Two elements $e, f \in E$ are called* **(anti-) parallel** *if $X_e = X_f$ $(X_e = -X_f)$ for every $X \in \mathcal{O}$. (Cf. Figure 8.1 above, in which e and ē are antiparallel.) $\mathcal{O}$ is said to be* **simple** *if E contains no loops or (anti-) parallel elements.*

Of course, if $e \in E$ is a loop then $\mathcal{O}$ and $\tilde{\mathcal{O}} := \mathcal{O} \setminus e$, considered as posets, are isomorphic. Similarly, if e and f are (anti-) parallel then $\mathcal{O}$ and $\tilde{\mathcal{O}} = \mathcal{O} \setminus e$ are isomorphic. Thus, since we are interested in OM posets only (rather than the OMs themselves), we will restrict our attention to simple OMs in what follows. The following result presents an alternative description of simple OMs which is intuitively obvious (cf. Figure 8.1):

Lemma 8.4 *Let $\mathcal{O}$ be an OM on E. Then the following are equivalent:*

(i) *$\mathcal{O}$ is simple.*

(ii) *If $e, f \in E$ and $e \neq f$, there exists $X \in \mathcal{O}$ such that $X_e = 0$ and $X_f \neq 0$.*

(iii) *If $[T]$ is a tope, then $supp\,T = E$. Moreover, every facet $[Y]$ of $[T]$ has a unique facet equator $e \in E$, i.e. $supp\,Y = E \setminus e$.*

Proof.

(i) $\Rightarrow$ (ii): Assume $\mathcal{O}$ is simple, and let $e, f \in E$ with $e \neq f$. Since f is not a loop, there exists $X' \in \mathcal{O}$ such that $X'_f \neq 0$. Since e and f are not parallel, we may choose X' such that $X'_e \neq X'_f$. If $X'_e = 0$, we are done. Thus assume $X'_e = -X'_f$. Since f is not a loop there exists $X'' \in \mathcal{O}$ such that $X''_f \neq 0$. Since e and f are not antiparallel, X'' may be chosen so that $X''_e \neq -X''_f$. If $X''_e = 0$, we are done. Thus assume $X''_e = X''_f$. Now eliminating e between X' and X'', fixing f, gives $X \in \mathcal{O}$ as claimed.

(ii) $\Rightarrow$ (iii): Assume (ii) and let $[T]$ be a tope of $\mathcal{O}$. Then $supp\,T = E$ is trivial. (Note that, by maximality of T, we get $T \circ X = T$ for every $X \in \mathcal{O}$.) Now let $[Y]$ be a facet of $[T]$ and assume it has two facet equators e and f, i.e. $supp\,Y \subseteq E \setminus \{e, f\}$. In fact, we may suppose that $supp\,Y = E \setminus \{e, f\}$, since if $g \in Y^0 \setminus \{e, f\}$, we may replace $\mathcal{O}$ by $\mathcal{O}/g$ (which again satisfies (ii)), leaving everything else unchanged. Now, by (ii), there exists $X \in \mathcal{O}$ such that $X_e = 0, X_f \neq 0$, say $X_f = T_f$. Then $Y' := Y \circ X$ shows that $[Y]$ cannot be a facet of $[T]$, a contradiction.

(iii) $\Rightarrow$ (i): Assume (iii) holds. If $e \in E$ then $e \in supp\,T$ for any tope $[T]$, hence E contains no loops. Next assume that e and f are parallel or antiparallel. Let $Y \in \mathcal{O}$ be maximal (with respect to $\preceq$) satisfying $Y_e = 0$. Since e and f are (anti-) parallel, this implies $Y_f = 0$, i.e. $supp\,Y \subseteq E \setminus \{e, f\}$. Again assume w.l.o.g. that $supp\,Y = E \setminus \{e, f\}$. If T covers Y in $\mathcal{O}$, then $T_e \neq 0$ and $T_f \neq 0$, since e and f are (anti-) parallel. Hence any such T has $supp\,T = E$, i.e. $[Y]$ is a facet of $[T]$ with two facet equators e and f, contradicting (iii).

□

A further simple observation is given by

Lemma 8.5 *Let $\mathcal{O}$ be simple. Let $[T]$ be a tope and let $e \in E = supp\,T$. Then the following are equivalent:*

(i) *e determines a facet* $[Y]$ *of* $[T]$.

(ii) *Both* $[T]$ *and* $[{}_{\bar{e}}T]$ *are topes.*

Thus, in particular, any facet is contained in precisely two topes.

Proof. This is trivial. (Note that ${}_{\bar{e}}T = Y \circ (-T)$ if and only if $supp\, Y = E \setminus e$.)

□

After these preliminary observations let us start with developing some tools for studying cells, or more generally "intervals" in OM-posets.

Definition 8.6 *Let* P *be a poset and let* $x \leq z$ *in* P. *Then*

$$[x, z] := \{y \in P \mid x \leq y \leq z\}$$

(endowed with the order induced from P*) is called an* **interval**.

The following lemma states that we may study tope lattices (which are special intervals $[T] = [0, T]$) instead of general intervals:

Lemma 8.7 (Isomorphism Lemma) *Every interval of an OM poset is isomorphic to some tope lattice. More precisely, let* $\mathcal{O}$ *be an OM, and let* $X \preceq Z$ *in* $\mathcal{O}$. *Let* $I := supp\, X$ *and let* $J := Z_0$. *Then*

$$\begin{array}{rcl} \varphi : [0, Z] \to \tilde{\mathcal{O}} & := & \mathcal{O}/I \setminus J \\ Y \to \tilde{Y} & := & Y \setminus (I \cup J) \end{array}$$

induces an isomorphism $\bar{\varphi} : [X, Z] \to [\tilde{Z}]$, *which is a tope in* $\tilde{\mathcal{O}}$.

Proof. φ is obviously order preserving and $\bar{\varphi}$, the restriction of φ to $[X, Z]$ is obviously injective. To prove that $\bar{\varphi}$ is surjective, let $\tilde{Y} \preceq \tilde{Z}$, and let $Y \in \varphi^{-1}(\tilde{Y})$. Then $X \circ Y \in [X, Z]$ is such that $\varphi(X \circ Y) = \tilde{Y}$. Hence $\tilde{Y} \in im\, \bar{\varphi}$.

□

As will become clear in subsequent sections, the structure of an OM poset is more or less determined by its "local structure", i.e. by the structure of intervals and its "global structure", i.e. the way in which the cells are put together in order to make up all of $\mathcal{O}$. The Isomorphism Lemma tells us that we may analyze the local structure of OM posets by considering topes. If $\mathcal{O}$ is an OM and we are interested in a tope $[T]$ of $\mathcal{O}$, then it is convenient to make a further simplification, which essentially consists in removing everything from E which does not affect $[T]$. More precisely, if $e \in E$ does not determine a facet of $[T]$, then we may replace $\mathcal{O}$ by $\tilde{\mathcal{O}} := \mathcal{O}/e$, leaving $[T]$ essentially unchanged, cf. Figure 8.2 below:

More precisely, this can be stated as follows:

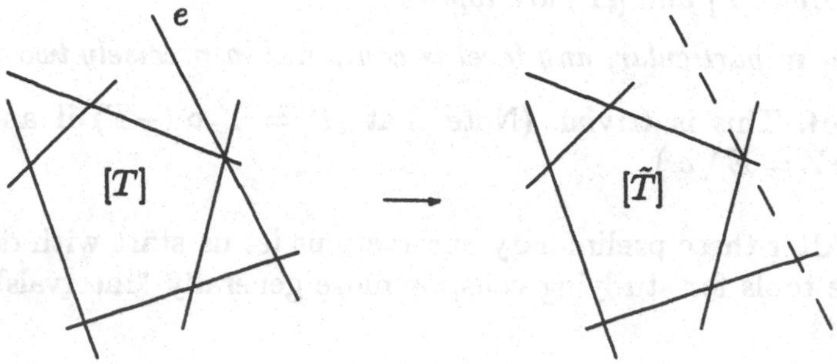

Figure 8.2

Lemma 8.8 *Let $\mathcal{O}$ be an OM on E, and let $[T]$ be a tope. If $e \in E$ does not determine a facet of $[T]$, then the map*

$$\begin{aligned} \varphi : \mathcal{O} &\rightarrow \mathcal{O}/e \\ X &\rightarrow \tilde{X} := X \setminus e \end{aligned}$$

induces an isomorphism $\bar{\varphi} : [T] \rightarrow [\tilde{T}]$.

Proof. Obviously, φ (and hence $\bar{\varphi}$) is order preserving. Thus we are left to show that $\bar{\varphi}$ is bijective. Let $[Y]$ be the face of $[T]$ determined by e, i.e. $[Y]$ is the unique maximal face of $[T]$ such that $Y_e = 0$.

(i) $\bar{\varphi}$ is surjective: Let $\tilde{X} \preceq \tilde{T}$. Since $\varphi : \mathcal{O} \rightarrow \tilde{\mathcal{O}}$ is surjective there exists $X \in \mathcal{O}$ such that $\varphi(X) = \tilde{X}$. We have to show that $X \preceq T$, i.e. $e \notin sep(X,T)$. Assume to the contrary that $e \in sep(X,T)$. Then $T' := {}_{\bar{e}}T = X \circ T \in \mathcal{O}$, and the approximation property of $\mathcal{O}$ yields that $Y = \tilde{T} + e^0 \in \mathcal{O}$, i.e. $[Y]$ is a facet of $[T]$ with facet equator e, contradicting our assumption.

(ii) φ is injective: Assume that $\bar{\varphi}(X) = \bar{\varphi}(Z)$ for $X, Z \in [T]$, say $X_e = 0$ and $Z_e = T_e \neq 0$. Then $T' := {}_{\bar{e}}T = X \circ (-Z) \circ T \in \mathcal{O}$ leading to a contradiction as in (i).

□

Definition 8.9 *Let $\mathcal{O}$ be a simple OM, and let $[T]$ be a tope of $\mathcal{O}$. Then $\mathcal{O}$ is said to be* **rooted** *at T if every $e \in E$ determines a facet of $[T]$.*

Thus, Lemma 8.8 above states that when studying a tope $[T]$ in $\mathcal{O}$, we may assume that $\mathcal{O}$ is rooted at T. This often simplifies matters considerably, as we will see in the following sections.

Let us complete our collection of "simplification tools" by mentioning the following obvious fact.

Lemma 8.10 *If $\mathcal{O}$ is rooted at T and $e \in E$, then $\tilde{\mathcal{O}} = \mathcal{O}/e$ is rooted at $\tilde{T} = T \setminus e$.*

Proof. This is trivial.

□

8.2 Basic Results

In this section we will use the tools developed in Section 8.1 in order to prove the existence of a dimension function, "path connectivity" of OM posets and some other basic results. Throughout, $\mathcal{O}$ will be assumed to be simple.

Lemma 8.11 *Every edge contains precisely two vertices.*

Proof. Let $[Z]$ be an edge, i.e. Z covers some elementary $X \in \mathcal{O}$. Then the claim states that the interval $[0, Z]$ contains precisely one more elementary vector $X' \neq X$. From the Composition Theorem, we conclude that there exists at least one such X'. So suppose there are two such vectors X' and X''. W.l.o.g. assume that $\mathcal{O}$ is simple and $Z \geq 0$. (Reorientation does not affect the poset $[Z]$.) Furthermore, by the Isomorphism Lemma 8.7, we may as sume that $[Z]$ is a tope. Finally, using Lemma 8.8, we may assume that $\mathcal{O}$ is rooted at Z. Hence $X = Z \setminus e$, where e is the (unique) facet equator of X. Now both X' and X'' must contain e in their support, since X is elementary and $supp\, X = E \setminus e$. Since X' and X'' are elementary, their supports are not contained in each other. Thus let $f \in supp\, X' \setminus supp\, X''$. Eliminate e between $-X'$ and X'', fixing f. The resulting Y has $Y_e = 0$, $Y_f = -$, and $Y \geq 0$ on $supp\, X'' \setminus supp\, X' \neq \emptyset$. Hence, in particular, $Y \neq \pm X$. On the other hand, $supp\, Y \subseteq E \setminus e = supp\, X$ together with the fact that X is elementary implies $Y = \pm X$, a contradiction.

□

Note that an OM poset $\mathcal{O}$ is in general not a lattice since it does not have a maximal element (except if $\mathcal{O} = \{0\}$). However, if we add such a maximal element $1_{\mathcal{O}}$ "artificially" then the enlarged set $\bar{\mathcal{O}} := \mathcal{O} \cup \{1_{\mathcal{O}}\}$ in fact becomes a lattice. Sometimes, results can be stated more elegantly using $\bar{\mathcal{O}}$ instead of $\mathcal{O}$. As an example, we will derive a generalization of the above result about edges.

Definition 8.12 *Let $X \preceq Z$ in $\hat{\mathcal{O}}$. If there exists a $Y \in \mathcal{O}$ such that Y covers X, and Z covers Y then $[X, Z]$ is called an* **interval of dimension 1**.

Theorem 8.13 *Every interval of dimension 1 in $\hat{\mathcal{O}}$ consists of precisely four elements. Its* HASSE *diagram looks like a diamond:*

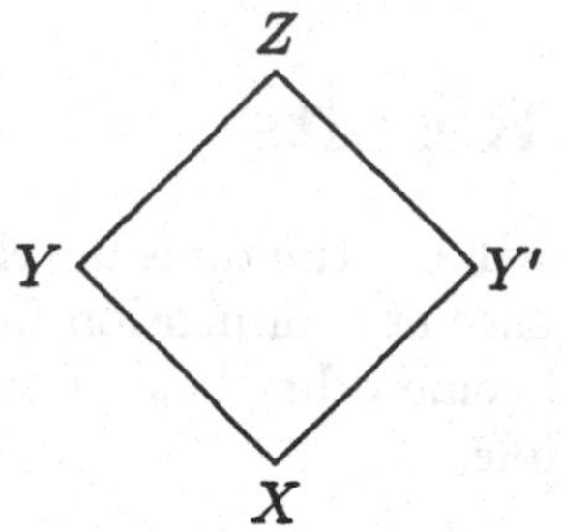

Proof. If $Z = 1_{\mathcal{O}}$, the claim states that every facet of a tope is contained in precisely two topes, which is the content of Lemma 8.5. If $Z \in \mathcal{O}$, then the Isomorphism Lemma 8.7 tells us that we may assume that $[X, Z] = [0, T] = [T]$ is a tope, i.e. $[T]$ is an edge. So in this case, the claim follows from Lemma 8.11 above.

□

Corollary 8.14 *Let $[T]$ be a tope, and let $[X, T]$ be an interval of dimension 1. If $\mathcal{O}$ is rooted at T then $supp\, X = E \setminus \{e_1, e_2\}$ for appropriate $e_1, e_2 \in E$.*

Proof. Assume that $supp\, X \subseteq E \setminus \{e_1, e_2, e_3\}$ for three different elements $e_1, e_2, e_3 \in E$. Since $\mathcal{O}$ is rooted at T, the e_i determine three facets $[Y^i]$ $(i = 1, 2, 3)$ of $[T]$, contradicting Theorem 8.13 above.

□

Our next goal is to show that OM-posets are "path connected" in the following sense: Recall from Lemma 8.5 that if $[T]$ is a tope, and $[Y]$ is a facet determined by $e \in E$ then (provided $\mathcal{O}$ is simple) $[T'] = [_{\bar{e}}T]$ is a tope again. Thus we can "move" from $[T]$ to $[T']$ by "switching" over a common facet. Path connectivity means that we can reach every tope by starting at $[T]$ and successively moving from one tope to another in this way.

Definition 8.15 *Two cells* $[X], [X']$ *are called* **adjacent**, *if they share a common facet. A sequence of cells* $[X^0], \ldots, [X^k]$ *is called a* **path** *from* $[X^0]$ *to* $[X^k]$, *if* $[X^{i-1}]$ *and* $[X^i]$ *are adjacent* $(i = 1, \ldots, k)$. *The number* k *is called the* **length** *of the path.*

Proposition 8.16 *If* $[T]$ *and* $[T']$ *are topes, then there exists a path from* $[T]$ *to* $[T']$ *of length* $k = |sep(T, T')|$.

Proof. The proof is by induction on $k = |sep(T, T')|$. If $k = 1$, the claim follows from Lemma 8.5. Thus let $k \geq 2$, and choose $e \in sep(T, T')$. Let $X \in \mathcal{O}$ be obtained by approximating T and T' on e. We may assume w.l.o.g. that $supp\, X = E \setminus e$. In fact, suppose that $X_f = 0$ for some $f \neq e$. Since $\mathcal{O}$ is simple, there exists $Y \in \mathcal{O}$ such that $Y_e = 0$, $Y_f \neq 0$ (cf. Lemma 8.4). Obviously, $X \circ Y$ is again an approximation of T and T' on e. This shows that we may in fact assume that $supp\, X = E \setminus e$. But then, by our inductive assumption, there exists a path from T to $X \circ T$, a path from $X \circ T$ to $X \circ T'$, and a path from $X \circ T'$ to T'. Concatenating these paths, we get one from T to T'. By induction, the length of this path equals

$$|sep(T, X \circ T)| + |sep(X \circ T, X \circ T')| + |sep(X \circ T', T')| = |sep(T, T')|.$$

□

Corollary 8.17 *If* $[T] \neq [T']$ *are topes, then there exists a facet* $[Y]$ *of* $[T]$ *with facet equator* $e \in sep(T, T')$.

Proof. Let $[T] = [T^0], [T^1], \ldots, [T^k] = [T']$ be a path as in the proof of Proposition 8.16 and choose $[Y] = [T] \cap [T^1]$.

□

The following result again concernes path connectivity:

Proposition 8.18 *Let* $[U]$ *be a cell, and let* $[Y]$ *and* $[Y']$ *be facets of* $[U]$. *Then there exists a path from* $[Y]$ *to* $[Y']$, *consisting of facets of* $[U]$.

Proof. By the Isomorphism Lemma 8.7, we may assume that $[U] = [T]$ is a tope. Furthermore, we may assume that $\mathcal{O}$ is simple and rooted at T. Finally, there is obviously no loss of generality if we assume that $T \geq 0$, i.e. $T = (+, \ldots, +)$. (In fact, reorientation induces an isomorphism of $\mathcal{O}$ as a poset.)

Let f, f' denote the facet equators of $[Y]$ and $[Y']$, resp. The proof is by induction on $|E|$. If $E = \{f, f'\}$, the claim is obviously true, since in this case $[Y]$ and $[Y']$ share $[0]$ as a common facet. Thus assume $e \in E \setminus \{f, f'\}$, and consider $\tilde{\mathcal{O}} := \mathcal{O}/e$ and the map $X \to \tilde{X} = X \setminus e$. By Lemma 8.10, $\tilde{\mathcal{O}}$ is rooted at $\tilde{T}$. By induction, there exists a path $[\tilde{Y}] = [\tilde{Y}^0], [\tilde{Y}^1], \ldots, [\tilde{Y}^k] = [\tilde{Y}']$ of facets of $[\tilde{T}]$, as indicated below.

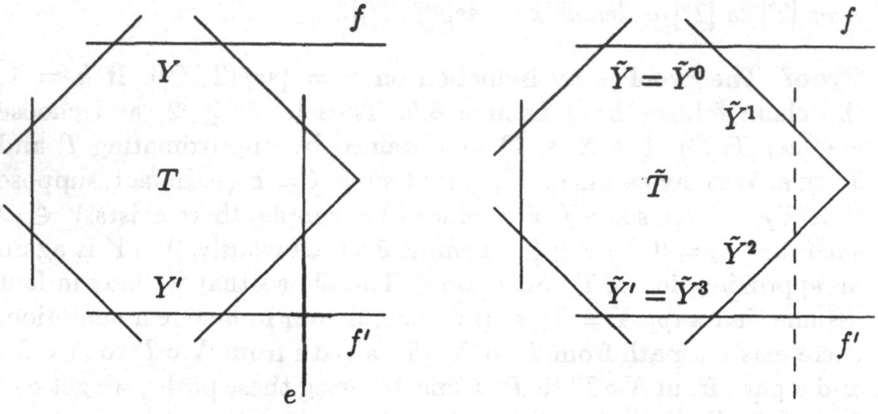

Then $Y^i := \tilde{Y}^i + e^+$ determines a sequence of facets of $[T]$ (but not necessarily a path). To construct a path from this sequence we do the following: Whenever $[Y^{i-1}]$ and $[Y^i]$ are not adjacent then we insert the facet $[X]$ of $[T]$, determined by e, between $[Y^{i-1}]$ and $[Y^i]$. We claim that the sequence arising this way is in fact a path. To see this, assume that $[Y^{i-1}]$ and $[Y^i]$ are not adjacent. Then we have to show that both are adjacent to $[X]$. Let f_{i-1} and f_i denote the facet equators of $[Y^{i-1}]$ and $[Y^i]$, resp. Then the situation is as follows (remember Corollary 8.14):

		f			f_{i-1}	f_i		f'	e		
Y^{i-1}	$=$	$+$	$\ldots$	$+$	0	$+$	$+\ldots$	$+$	$+$	$\in \mathcal{O}$	
Y^i	$=$	$+$	$\ldots$	$+$	$+$	0	$+\ldots$	$+$	$+$	$\in \mathcal{O}$	
X	$=$	$+$	$\ldots$	$+$	$+$	$+$	$+\ldots$	$+$	0	$\in \mathcal{O}$	
Z	$=$	$+$	$\ldots$	$+$	0	0	$+\ldots$	$+$	$+$	$\notin \mathcal{O}$,	but
$\tilde{Z}$	$=$	$+$	$\ldots$	$+$	0	0	$+\ldots$	$+$		$\in \tilde{\mathcal{O}}$,	hence
Z'	$=$	$+$	$\ldots$	$+$	0	0	$+\ldots$	$+$	$\ominus$	$\in \mathcal{O}$.	

Suppose first that $Z'_e = 0$. Obviously, the interval $[Z', Y^{i-1}]$ can not be of dimension 1 (cf. Theorem 8.13), since $Z \notin \mathcal{O}$. Hence Y^{i-1} covers Z' (and, similarly, Y^i covers Z'). But then $[Z', T]$ is an interval of dimension 1 containing at least five elements Z', Y^{i-1}, Y^i, X and T, which is impossible. Thus, suppose that $Z'_e = -$. In this case, approximating Y^{i-1} and Z' on e, we get a common facet of Y^{i-1} and X. Thus Y^{i-1} and X are adjacent. Similarly, Y^i and X are adjacent, which proves the claim.

□

Theorem 8.19 *Let* $(\mathcal{O}, \preceq)$ *be an OM poset. Then* $\mathcal{O}$ *is JD, i.e. there exists a function* $\dim : \mathcal{O} \to \mathbb{Z}$ *such that* $\dim 0 = -1$ *and* $\dim Y = \dim X + 1$ *whenever* Y *covers* X *in* $\mathcal{O}$.

Proof. We have to show that, given $X \in \mathcal{O}$, all maximal chains from 0 to X have the same length. The proof is by induction on the maximum length k of a maximal chain from 0 to X. For $k \leq 1$ there is nothing to show. The case $k = 2$ is settled by Theorem 8.13. Thus suppose that $k \geq 3$. Choose any maximal chain of length k, say $0 = Z^1 \prec\!\!\cdot \ldots \prec\!\!\cdot Z^k \prec\!\!\cdot X$. We have to show that every other maximal chain $0 = Y^1 \prec\!\!\cdot Y^2 \prec\!\!\cdot \ldots \prec\!\!\cdot Y^m \prec\!\!\cdot X$ has the same length, i.e. $k = m$. First observe that both $[Z^k]$ and $[Y^m]$ are facets of $[X]$. By Proposition 8.18 there exists a path $[Z^k] = [U^0], [U^1], \ldots, [U^n] = [Y^m]$ from $[Z^k]$ to $[Y^m]$, cf. Figure 8.3 below.

From this it is obvious, how to complete the proof by choosing maximal chains from 0 to all the common facets of $[U^{i-1}]$ and $[U^i]$, successively concluding by induction that the "dimension" of $Z^k = U^0$ is the same as that of $U^1, U^2, \ldots, U^n = Y^m$.

□

Definition 8.20 *Let* $\dim : \mathcal{O} \to \mathbb{Z}$ *as in Theorem 8.19. Then* $\dim X$ *is called the* **dimension** *of* $X \in \mathcal{O}$. $\dim \mathcal{O} := \max\{\dim X \mid X \in \mathcal{O}\}$ *is called the* **dimension** *of* $\mathcal{O}$.

Theorem 8.21 *All topes of* $\mathcal{O}$ *have the same dimension* $\dim \mathcal{O}$.

Proof. This is proved in exactly the same way as Theorem 8.19, this times using path connectivity of topes (rather than facets), as provided by Proposition 8.16.

□

Theorems 8.19 and 8.21 may be summarized by saying that $\hat{\mathcal{O}} = \mathcal{O} \cup \{1_{\mathcal{O}}\}$ has a dimension function. We may also summarize our two results on path connectivity:

Theorem 8.22 *Any two cells of the same dimension are connected by a path. (Of course, conversely, any two cells connected by a path must have the same dimension.)*

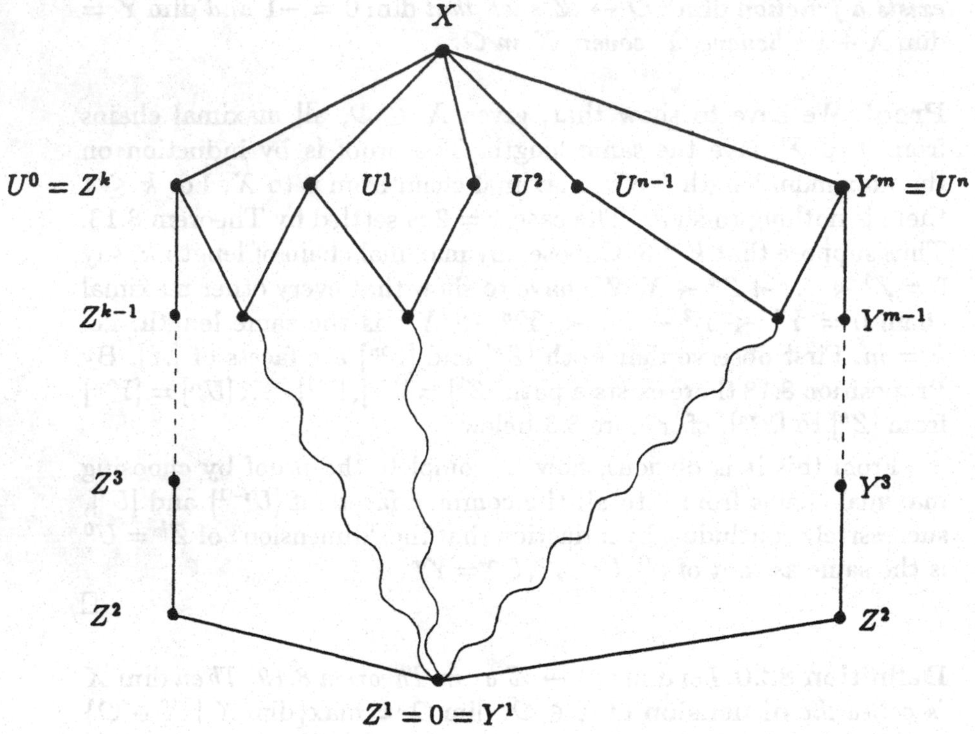

Figure 8.3

Proof. Let $[X]$ and $[X']$ have equal dimension d. The proof is by induction on $r := \dim \mathcal{O} - d$. If $r = 0$, then $[X]$ and $[X']$ are topes and hence they are connected by a path due to Proposition 8.16. Thus let $r \geq 1$, and assume inductively that the claim holds for $r-1$. Choose $Y, Y' \in \mathcal{O}$ covering X and X'. resp., and let $[Y] = [Y^0], [Y^1], \ldots, [Y^k] = [Y']$ be a path. Let $[X^i]$ denote the common facet of $[Y^{i-1}]$ and $[Y^i]$ $(i = 1, \ldots, k)$, and let $[X^0] = [X]$ and $[X^{k+1}] := [X']$. Then for $i = 1, \ldots, k+1$, $[X^{i-1}]$ and $X^i]$ are both facets of $[Y^{i-1}]$, hence they are connected by a path due to Proposition 8.18. Concatenating these paths gives a path from $[X]$ to $[X']$.

□

Just for the sake of completeness, let us present one more property of OM posets. This will not be used in what follows (and hence may be skipped without loss in the first reading).

Definition 8.23 *A lattice* $(L, \leq)$ *is called* **complemented** *if for every* $x \in L$ *there exists an* $\bar{x} \in L$ *such that* $x \wedge \bar{x} = 0$ *and* $x \vee \bar{x} = 1$. *L is called* **relatively complemented** *if every interval of L is complemented.*

Proposition 8.24 $\bar{\mathcal{O}}$ *is relatively complemented. Moreover, if* $[X, Z]$ *is an interval and* $Y \in [X, Z]$*, then the complement* $\bar{Y}$ *of* Y *in* $[X, Z]$ *can be chosen such that* $\dim Y + \dim \bar{Y} = \dim X + \dim Z$. *(Recall that* $\dim : \mathcal{O} \to \mathbb{Z}$ *can be extended to* $\dim : \bar{\mathcal{O}} \to \mathbb{Z}$.*)*

Proof. Let $[X, Z]$ be an interval and let $Y \in [X, Z]$. If $[X, Z]$ is an interval of dimension 1, the claim follows immediately from Theorem 8.13. We proceed by induction on $k := \dim Z - \dim X$. If $k \geq 3$ then either one of the following happens:

1) Y is not covering X, or

2) Z is not covering Y.

Suppose first that 1) holds. Let $U \in [X, Y]$ cover X. By induction, there exists a complement $\bar{Y}$ of Y in the interval $[U, Z]$, and a complement $\bar{U}$ of U in the interval $[X, \bar{Y}]$, cf. Figure 8.4 below.

We claim that $\bar{U}$ is a complement of Y in $[X, Z]$. In fact,

$$\bar{U} \vee Y \geq \bar{U} \vee U = \bar{Y}. \text{ Hence } \bar{U} \vee Y \geq \bar{Y} \vee Y = Z.$$

Similarly,

$$\bar{U} \wedge Y \leq \bar{Y} \wedge Y = U. \text{ Hence } \bar{U} \wedge Y \leq \bar{U} \wedge U = X.$$

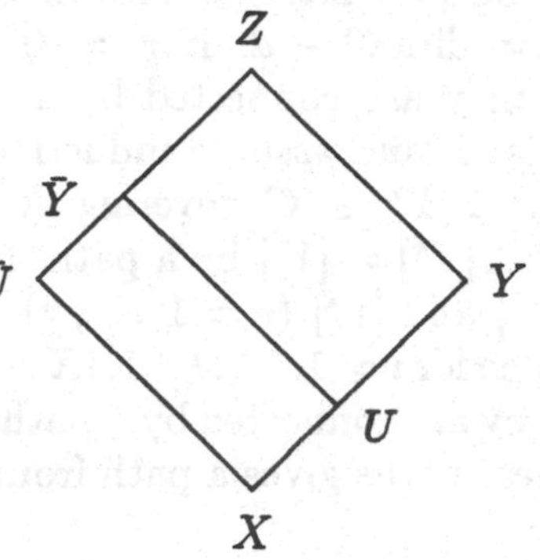

Figure 8.4

Since $\bar{U}, Y \in [X, Z]$, this implies that actually equality holds, i.e. $\bar{U} \vee Y = Z$ and $\bar{U} \wedge Y = X$. The dimension formula also follows immediately by induction. Case 2) is treated similarly.

□

8.3 Shellability of Topes

In this section, we will study tope lattices in more detail. By now, all we know about tope lattices is that they are JD (thus allowing a dimension function $\dim : [T] \to \mathbb{Z}$), that their intervals of dimension 1 look like a diamond and that tope lattices are relatively complemented. Since every facet $[Y]$ of a tope $[T]$ is again (isomorphic to) a tope lattice by Lemma 8.7, the structure of $[T]$ is more or less fully determined, once we know how the facets of $[T]$ are put together to make up all of $[T]$. This will in fact be our main concern in this section. We will show that $[T]$ can be "constructed" in a certain way by pasting together its facets, one by one, according to some ordering $[Y^1], \ldots, [Y^k]$.

To give an intuitive description of what we have in mind, think of $[T]$ as being the face lattice of a 3-dimensional polytope. Thus, topologically, $[T]$ corresponds to a 3-dimensional ball while every facet $[Y]$ of $[T]$ is a 2-dimensional ball. The union of all facets of $[T]$ corresponds to the boundary of the polytope, i.e. to a 2-dimensional sphere. We intend to show that the boundary of the polytope can be "constructed" by starting with any facet, say $[Y^1]$, and then adding successively all other facets $[Y^2], [Y^3], \ldots, [Y^k]$, one by one, such that in every intermediate step $i < k$ the union $[Y^1] \cup \ldots \cup [Y^i]$ is a 2-dimensional ball. To put it the other way round, we will prove that the boundary of the polytope can be "shelled" (or pealed) in the sense that we can remove all facets from the boundary, one by one, according to some prescribed ordering $[Y^k], [Y^{k-1}], \ldots, [Y^1]$, maintaining a topological 2-dimensional ball in the intermediate steps of the shelling process.

As it will become clear in chapter 9, the important feature of the above construction process is not that we can construct the boundary of $[T]$ by adding one cell at a time. Rather, the crucial point is that we can "construct" the boundary step by step, starting with some part of the boundary (possibly consisting of several facets), which is a topological 2-dimensional ball, and then, in each step, adding another part of the boundary (again, possibly consisting of several facets), which is a topological 2-ball, in such a way, that, what we get in each intermediate step, is a topological 2-ball again. This more general kind of "constructivity" will be described formally in terms of posets below. A similar formal definition of "shellability" will follow, since, according to our intuitive description of the construction process above, we will prove constructibility of tope lattices and OM-posets by actually showing that they are "shellable".

To begin with, let us translate the above topological notions of "balls", "spheres", and "boundary" into the context of posets.

Definition 8.25 *Let P be a poset, and let $Q \subseteq P$ be closed, i.e. $Q = [Q]$. Then*

$$\partial Q := \{q' \in Q \mid q' \leq q \text{ for some } q \in Q \text{ which is covered by exactly one maximal element of } Q\}$$

is called the **boundary** *of Q.*

Example 8.26 *If P has a dimension function* dim : $P \to \mathbb{Z}$ *and* $\dim Q = n$ *(i.e. $n = \max\{\dim q \mid q \in Q\}$), then* $\dim(\partial Q) \leq n - 1$. *If Q has a unique maximal element q then $\partial Q = Q \setminus q$.*

If $\mathcal{O}$ is an OM poset then $\partial \mathcal{O} = \emptyset$. If $[Z]$ is a cell then $\partial[Z] = [Z] \setminus \{Z\}$. From Theorem 8.13 we conclude that $\partial(\partial[Z]) = \emptyset$.

The notions of "balls" and "spheres" mentioned in the introduction will be replaced by the following notions of "B-constructivity" and "S-constructivity":

Definition 8.27 ("Constructivity")
Let P be a pure n-dimensional poset, i.e. there exists a dimension function dim : $P \to \mathbb{Z}$ *and all maximal elements of P have the same dimension n.*
*P is called S-***constructible** *of dimension n if either $n = -1$ or*

$P = P_1 \cup P_2$, with P_1 and P_2 being two B-constructible order ideals of dimension n such that $P_1 \cap P_2 = \partial P_1 = \partial P_2$ is S-constructible of dimension $n-1$.
*P is called B-***constructible** *of dimension n if either P has a unique maximal element p, i.e. $P = [p]$, and $\partial[p]$ is S-constructible of dimension $n-1$ or $P = P_1 \cup P_2$ with P_1 and P_2 being two B-constructible order ideals of dimension n such that $P_1 \cap P_2 = \partial P_1 \cap \partial P_2$ is B-constructible of dimension $n-1$.*

Example 8.28 *We present posets by means of their* HASSE *diagrams:*

is S-constructible of dimension -1.

is B-constructible of dimension 0.

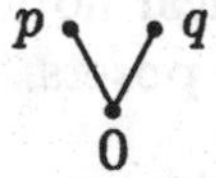

is S-constructible of dimension 0.

is not constructible.

is B-constructible of dimension 1.

is B-constructible of dimension 1.

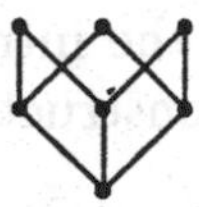
is S-constructible of dimension 1.

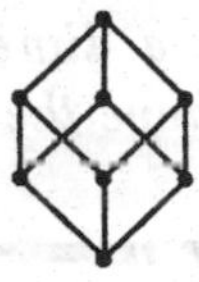
is B-constructible of dimension 2.

Example 8.29 *Consider the face lattices of the following polytopes:*

polytope	*figure*	*face lattice*
$P = \{p\}$:		
$P = conv\{p, q\}$:	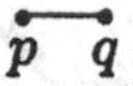	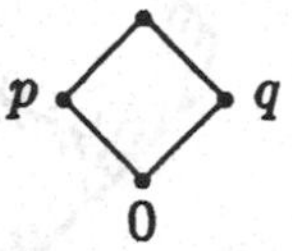
$P = conv\{p, q, r\}$:	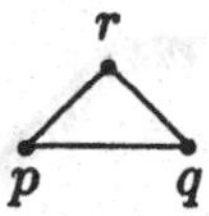	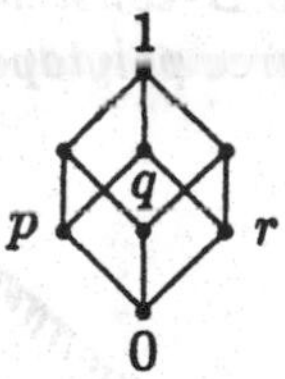

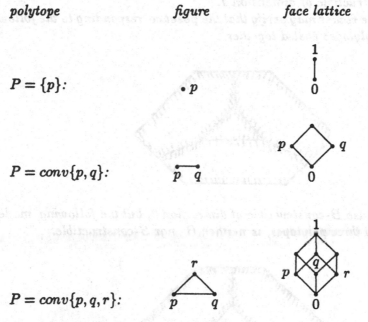

These are B-constructible of dimension 0,1, and 2, resp. Furthermore, consider the following two polytopes having a face in common:

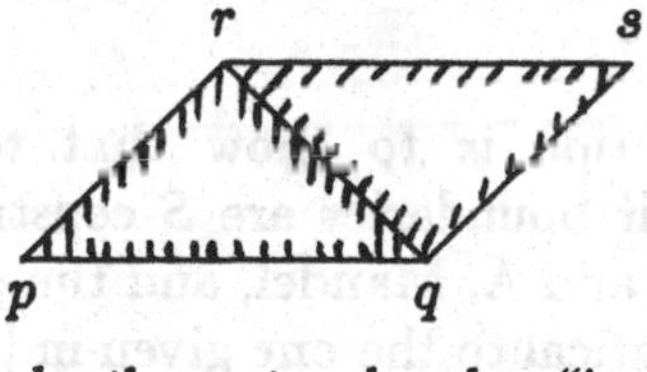

The poset arising by the natural order "is a facet of" is given by the HASSE *diagram*

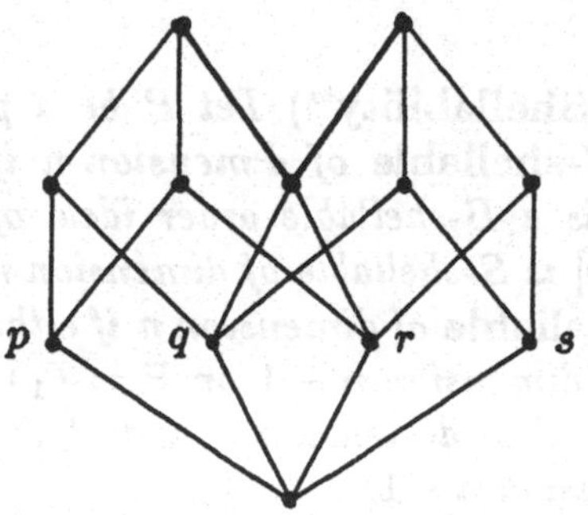

This is also B-constructible of dimension 2, and its boundary is S-constructible of dimension 1.

The reader may verify that the poset corresponding to the following four polytopes pasted together

is also B-constructible of dimension 2, but the following, made up of only three polytopes, is neither B- nor S-constructible.

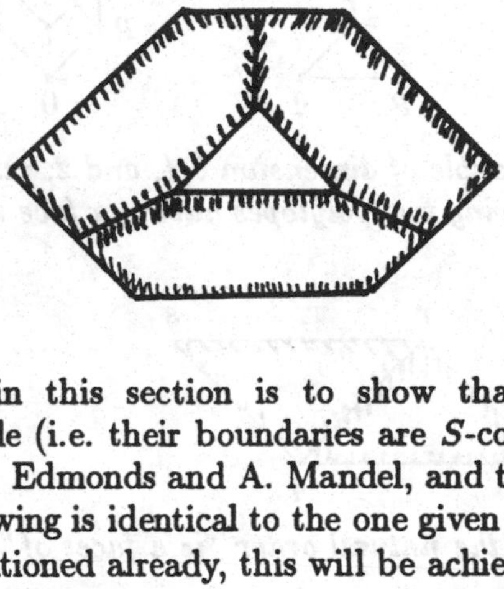

Our goal in this section is to show that tope lattices are B-constructible (i.e. their boundaries are S-constructible). This result is due to J. Edmonds and A. Mandel, and the proof we will present in the following is identical to the one given in [126].

As mentioned already, this will be achieved by showing that the boundary of a tope can be constructed by adding a single facet in each step. Thus we will actually prove that tope lattices are shellable in the following sense:

Definition 8.30 ("Shellability") *Let P be a pure n-dimensional poset. P is called S-***shellable** *of dimension n if $n = -1$ or $P = P_1 \cup [p]$, where P_1 is a B-shellable order ideal of dimension n and $P_1 \cap [p] = \partial P_1 = \partial[p]$ is S-shellable of dimension $n-1$.*

*P is called B-***shellable** *of dimension n if either $P = [p]$ with $\partial[p]$ being S-shellable of dimension $n-1$ or $P = P_1 \cup [p]$, where P_1 is a B-shellable order ideal of dimension n and $P_1 \cap [p] = \partial P_1 \cap \partial[p]$ is B-shellable of dimension $n-1$.*

Obviously, every shellable poset is also constructible. However, there do exist examples of constructible posets which are not shellable, cf. M.E. Rudin [145]. The only reason for introducing shellability here is that the most natural way to prove constructibility of tope lattices consists in actually showing that they are shellable. The following notion of "umbrellas" will play a central role in the proof of shellability of tope lattices.

Definition 8.31 *Let $Z \in \mathcal{O}$. Then a collection $\mathcal{U} \neq \emptyset$ of facets of $[Z]$ is called an* **umbrella** *of $[Z]$ if there exists $Z' \in \mathcal{O}$ such that $\operatorname{supp} Z = \operatorname{supp} Z'$ and $\mathcal{U}$ is the set of facets $[Y]$ of $[Z]$ such that $[Y]$ has a facet equator separating Z and Z'. $\mathcal{U}$ is then denoted by $\mathcal{U}(Z, Z')$. An umbrella of $[Z]$ is called* **proper** *if it does not contain all facets of $[Z]$. Figure 8.5 shows an umbrella of a tope $[T]$.*

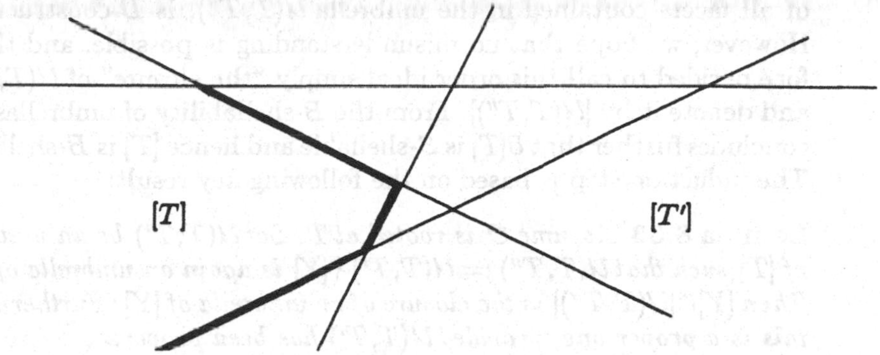

Figure 8.5

The above Definition 8.31 of umbrellas is in a sense unnecessarily complicated. In fact, we could obviously have restricted ourselves to considering umbrellas of topes, since if Z and Z' are as in Definition 8.31, then $[Z \setminus I]$ and $[Z' \setminus I]$ are topes in $\mathcal{O} \setminus I$ where $I = Z^0 = (Z')^0$. Now, if $[T]$ is a tope, we may assume that $\mathcal{O}$ is rooted at T and then $\mathcal{U}(T, T')$ is simply the set of facets of $[T]$ determined by the elements $e \in \mathit{sep}(T, T')$. However, we will have to consider both umbrellas of a tope $[T]$ and of its facets $[Y]$. So if umbrellas were defined only for topes, this would imply that we have to work in different OMs $\mathcal{O}$ and $\mathcal{O} \setminus I$ simultaneously, leading to some notational inconveniences.

We also would like to note that umbrellas could have been defined by means of paths as well: Suppose, for example, that $[T]$ and $[T']$ are topes. Then $\mathcal{U}(T, T')$ is precisely the set of facets of $[T]$ whose facet equators are "traversed" when moving from $[T]$ to $[T']$ along a path. From this the following is obvious.

Lemma 8.32 *Let $[T]$ be a tope, and let $\mathcal{U}(T,T')$ be an umbrella of $[T]$ containing at least two facets. Then there exists a facet $[Y] \in \mathcal{U}(T,T')$ and a tope $[T'']$ of $\mathcal{O}$ such that $\mathcal{U}(T,T'') = \mathcal{U}(T,T') \setminus [Y]$ is an umbrella again.*

Proof. Let $[T^0], \ldots, [T^k]$ be a path from $[T]$ to $[T']$, as constructed in Proposition 8.16 and let i be the last index such that $sep(T^{i-1}, T^i)$ determines a facet of $[T]$. (Note that if $sep(T^{i-1}, T^i)$ contains two or more elements, then these are parallel or antiparallel, hence determine the same facet.) Then $T'' := T^{i-1}$ is what we are looking for.

□

The notion of umbrellas already indicates how we are going to prove shellability of topes. First, we will show that the closure $[\mathcal{U}(T,T')]$ of a proper umbrella $\mathcal{U}(T,T')$ of $[T]$ is B-shellable. More precisely, we should say that the order ideal, consisting of the union of all facets contained in the umbrella $\mathcal{U}(T,T'')$, is B-constructible. However, we hope that no misunderstanding is possible, and therefore decided to call this order ideal simply "the closure" of $\mathcal{U}(T,T'')$, and denote it by $[\mathcal{U}(T,T'')]$. From the B-shellability of umbrellas one concludes further that $\partial[T]$ is S-shellable and hence $[T]$ is B-shellable. The induction step is based on the following key result:

Lemma 8.33 *Assume $\mathcal{O}$ is rooted at T. Let $\mathcal{U}(T,T')$ be an umbrella of $[T]$ such that $\mathcal{U}(T,T'') := \mathcal{U}(T,T') \setminus [Y]$ is again an umbrella of $[T]$. Then $[Y] \cap [\mathcal{U}(T,T'')]$ is the closure of an umbrella of $[Y]$. Furthermore, this is a proper one, provided $\mathcal{U}(T,T')$ has been proper.*

Proof. Let e denote the facet equator of $[Y] = \mathcal{U}(T,T') \setminus \mathcal{U}(T,T'')$. Thus $e = sep(T', T'')$. Let $[Y'']$ denote the common facet of $[T']$ and $[T'']$. Then the facet equator of $[Y'']$ is also e, i.e. $supp\, Y = supp\, Y''$ and hence $\mathcal{U}(Y,Y'')$ is a well defined umbrella of $[Y]$. We show that $[Y] \cap [\mathcal{U}(T,T'')] = [\mathcal{U}(Y,Y'')]$, which proves the first claim.

"$\supseteq$": It is sufficient to show that $\mathcal{U}(Y,Y'') \subseteq [Y] \cap [\mathcal{U}(T,T'')]$. Thus let $[X] \in \mathcal{U}(Y,Y'')$ be a facet of $[Y]$ with facet equator, say, f. Thus $f \in sep(Y,Y'')$. Let $[Y']$ denote the facet of $[T]$ determined by f. Thus $[X] \subseteq [Y']$, so it suffices to show that $[Y'] \in \mathcal{U}(T,T'')$, i.e. $f \in sep(T,T'')$. But this is clear since $f \in sep(Y,Y'')$.

"$\subseteq$": Let $Z \in [Y] \cap [\mathcal{U}(T,T'')]$. Then $Z_f = 0$ for some $f \in sep(T,T'') = sep(T,T') \setminus e$. Let $Y' := Z \circ Y''$. Thus $Y \neq Y'$ and $\mathcal{U}(Y,Y') \subseteq \mathcal{U}(Y,Y'')$. We show that $Z \in [\mathcal{U}(Y,Y')]$. Take any facet $[X] \in \mathcal{U}(Y,Y')$. Then the facet equator g of $[X]$ is in $sep(Y,Y')$, hence $g \notin supp\, Z$. Thus $Z \preceq X$, showing that $[Z] \in [\mathcal{U}(Y,Y')]$.

Thus, the first claim is settled. Now suppose that $\mathcal{U}(T,T')$ is proper but $\mathcal{U}(Y,Y'')$ is not, i.e. $[Y] \cap [\mathcal{U}(T,T'')] = [\mathcal{U}(Y,Y'')] = \partial[Y]$, i.e. $\partial[Y] \subseteq [\mathcal{U}(T,T'')]$. Then the following lemma shows that either $[Y] \in \mathcal{U}(T,T'')$ or ${}_{\bar{e}}T'' = -T$, implying that $T' = -T$, both contradicting our assumption. Thus the second claim also holds.

□

Lemma 8.34 *Let $\mathcal{O}$ be rooted at T, and let $\mathcal{U}(T,T'')$ be an umbrella such that $\partial[Y] \subseteq [\mathcal{U}(T,T'')]$ for some facet $[Y]$ of $[T]$ with facet equator e. Then either ${}_{\bar{e}}T'' = -T$ or $[Y] \in \mathcal{U}(T,T'')$.*

Proof. The proof is by induction on $|E|$. If $|E| = 1$, the claim is trivial. So assume $|E| \geq 2$. If ${}_{\bar{e}}T'' \neq -T$, there exists $f \neq e$ which is not separating T and T''. Now consider $\tilde{\mathcal{O}} = \mathcal{O}/f$ and the map $X \rightarrow \tilde{X} = X \setminus f$. By Lemma 8.10, $\tilde{\mathcal{O}}$ is rooted at $\tilde{T}$. Furthermore, $[\tilde{Y}]$ is a face of $[\tilde{T}]$ and $\partial[\tilde{Y}] \subseteq [\mathcal{U}(\tilde{T},\tilde{T}'')]$. Hence, by induction, either ${}_{\bar{e}}\tilde{T}'' = -\tilde{T}$, implying that ${}_{\bar{e}}T'' = -T$ (since f is not separating T and T''), or $[\tilde{Y}] \subseteq [\mathcal{U}(\tilde{T},\tilde{T}'')]$, i.e. $e \in sep(\tilde{T},\tilde{T}'')$ and hence $[Y] \subseteq [\mathcal{U}(T,T'')]$.

□

Theorem 8.35 *Let $[T]$ be a tope of dimension n. Then*

(i) *The closure of every proper umbrella of $[T]$ is B-shellable of dimension $n-1$.*

(ii) *$\partial[T]$ is S-shellable of dimension $n-1$.*

(iii) *$[T]$ is B-shellable of dimension n.*

Proof. The proof is by induction on n. If $n = 0$, the claims are trivial. If $n = 1$, they follow from Theorem 8.13. Thus assume $n \geq 2$, and that $\mathcal{O}$ is rooted at T.

ad (i): Let $\mathcal{U}$ be a proper umbrella of $[T]$. If $\mathcal{U}$ consists of a single facet $[Y]$ of $[T]$, then $[Y]$ is (isomorphic to) an $(n-1)$-dimensional tope, thus it is B-shellable of dimension $n-1$ by induction. We proceed by induction on $k := |\mathcal{U}|$. If $k \geq 2$, there exists a facet $[Y] \in \mathcal{U}$ such that $\mathcal{U} \setminus [Y]$ is a proper umbrella (by Lemma 8.32), hence its closure is B-shellable of dimension $n-1$ by induction on k. $[Y]$ itself is B-shellable of dimension $n-1$, and Lemma 8.33 implies that $[Y] \cap [\mathcal{U} \setminus [Y]]$ is B-shellable of dimension $n-2$. Thus $[\mathcal{U}]$ is B-shellable of dimension $n-1$ by definition of B-shellability.

ad (ii): Let $[Y]$ be a facet of $[T]$, and let e be its facet equator. Then $[\partial[T] \setminus [Y]] = [\mathcal{U}(T, -_e T)]$ is B-shellable of dimension $n-1$, and $[Y]$ is also B-shellable of dimension $n-1$. Their intersection is $[\partial[T] \setminus [Y]] \cap [Y] = \partial[Y]$ which, by induction, is S-shellable of dimension $n-2$. Hence $\partial[T] = [\partial[T] \setminus [Y]] \cup [Y]$ is S-shellable of dimension $n-1$.

(iii) is trivial by definition of B-shellability.

□

8.4 Constructibility of $\mathcal{O}$

In this section we will show that every OM-poset $\mathcal{O}$ is S-constructible, i.e. that $\hat{\mathcal{O}} = \mathcal{O} \cup 1_{\mathcal{O}}$ is B-constructible. Again our proof will consist in actually showing that $\mathcal{O}$ is shellable. This result has been independently obtained by Lawrence [123] and Hochstättler [102] .

Definition 8.36 *Let $\mathcal{O}$ be an OM-poset of dimension n, corresponding to a simple OM on some set E. Let $[T]$ be a tope of $\mathcal{O}$, let r be a nonnegative integer and let $\mathcal{L}$ be a set of topes such that*

(i) $r = \max_{[T'] \in \mathcal{L}} |sep(T, T')|$

(ii) *$\mathcal{L}$ contains all topes $[T']$ with $|sep(T, T')| < r$.*

Then $\mathcal{L}$ is called a **lump** *(centered at T of radius r).*

Our object is to show that a lump is always B-shellable unless it contains all topes of $\mathcal{O}$. From this, shellability of OM-posets will then be straightforward.

Lemma 8.37 *Let $\mathcal{L}$ be a lump centered at T of radius $r \geq 1$. Then there exists a tope $[T'] \in \mathcal{L}$ such that $\mathcal{L}' := \mathcal{L} \setminus \{T'\}$ is again a lump. Furthermore, we have*

$$[\mathcal{L}'] \cap [T'] = \partial[\mathcal{L}'] \cap \partial[T'] = [\mathcal{U}(T', T)] .$$

(Here, of course, $[\mathcal{L}']$, the "closure" of $\mathcal{L}$, is to be understood in the same way as the "closure" of an umbrella.)

Proof. The first claim is obvious. In fact, removing any tope $T' \in \mathcal{L}$ with $sep(T, T') = r$ leaves a lump $\mathcal{L}' := \mathcal{L} \setminus [T']$ of radius r or $r-1$.

Now let us show that if T' and $\mathcal{L}'$ are as above, then

$$[\mathcal{L}'] \cap [T'] = \partial[\mathcal{L}'] \cap \partial[T']$$

"$\supseteq$" is trivial. To prove the converse inclusion, let $Z \in \lceil\mathcal{L}'\rceil \cap \lceil T'\rceil$. Hence $Z \preceq T'$ and $Z \preceq T''$ for some tope $T'' \in \mathcal{L}$. Since $sep(T,T')$ and $sep(T,T'')$ are both less or equal to r, we obviously have $sep(Z,T) =: i < r$. Thus $S := Z \circ T \in \mathcal{L}'$. Choose any path from T to S and append any path from S to T', both paths being constructed as in Proposition 8.16. This results in a path

$$T = T_0, T_1, \ldots, T_i = S, T_{i+1}, \ldots, T_r = T' .$$

Note that $|sep(T,T_j)| = j$ for all j, hence $T_1, \ldots, T_{r-1} \in \mathcal{L}'$. Along the path from T to S we "traverse" all elements $e \in sep(S,T)$. (We assume that $\mathcal{O}$ is simple, hence we traverse exactly one element $e \in E$ at a time.) Along the path from S to T' we only traverse all elements in $sep(T,T') \setminus sep(T,S)$. From this it is clear that Z conforms to all topes $T_i, T_{i+1}, \ldots, T_r$. In particular $Z \preceq T_{r-1}$. Thus, if $[Y]$ denotes the common facet of T_{r-1} and $T_r = T'$, then $Z \in [Y]$. But $[Y]$ obviously is an element of $\partial\lceil\mathcal{L}'\rceil$ and $\partial\lceil T'\rceil$ (recall that $T_{r-1} \in \mathcal{L}'$). Hence $Z \in \partial\lceil\mathcal{L}'\rceil \cap \partial\lceil T'\rceil$, which proves the inverse inclusion.

Note that our above argument actually shows that if $[Z] \subseteq \lceil\mathcal{L}'\rceil \cap \lceil T'\rceil$, then $[Z] \subseteq [Y]$, where $[Y]$ is a facet of $\lceil T'\rceil$ with facet equator $e \in sep(T',T)$, i.e. $[Y] \in \mathcal{U}(T',T)$. Hence $\lceil\mathcal{L}'\rceil \cap \lceil T'\rceil \subseteq [\mathcal{U}(T',T)]$. The converse inclusion is clear, since every facet $[Y] \subseteq [\mathcal{U}(T',T)]$ has a facet equator $e \in sep(T',T)$. Hence ${}_{\bar{e}}T' \in \mathcal{L}'$, showing that $[Y] \in \lceil\mathcal{L}'\rceil \cap \lceil T'\rceil$. Thus we get

$$\lceil\mathcal{L}'\rceil \cap \lceil T'\rceil = [\mathcal{U}(T',T)]$$

□

Theorem 8.38 *Let $\mathcal{O}$ be an OM-poset of dimension n. Then*

(i) *If $\mathcal{L}$ is a lump of radius $r < |E|$, then $\lceil\mathcal{L}\rceil$ is B-shellable of dimension n.*

(ii) *$\mathcal{O}$ is S-shellable of dimension n.*

(iii) *$\hat{\mathcal{O}} = \mathcal{O} \cup \{1_{\mathcal{O}}\}$ is B-shellable of dimension $n+1$.*

Proof. (i): Let $\mathcal{L}$ be a lump centered at T of radius $r < |E|$ (i.e. $-T \notin \mathcal{L}$). If $|\mathcal{L}| = 1$, then $\lceil\mathcal{L}\rceil = \lceil T\rceil$ is B-shellable of dimension n by Theorem 8.35. If $|\mathcal{L}| \geq 2$, then, by Lemma 8.37, we find $T' \in \mathcal{L}$ such that $\mathcal{L}' = \mathcal{L} \setminus \lceil T'\rceil$ is a lump and

$$\lceil\mathcal{L}'\rceil \cap \lceil T'\rceil = \partial\lceil\mathcal{L}'\rceil \cap \partial\lceil T'\rceil = [\mathcal{U}(T',T)]$$

which is B-shellable of dimension $n-1$ by Theorem 8.37, because $\mathcal{U}(T', T)$ is a proper umbrella of T'. (Note that $T' \neq -T$, since $[-T] \notin \mathcal{L}$.) Hence we conclude inductively that $\mathcal{L}$ is B-shellable.

(ii): Let $[T]$ be a tope and let $\mathcal{L}$ consist of all topes, i.e. $\mathcal{L}$ is the lump centered at $[T]$ of radius $r = |E|$. Let $\mathcal{L}' = \mathcal{L} \setminus [-T]$ (which is the unique lump that can be obtained from $\mathcal{L}$ by removing a tope $[T']$). From Lemma 8.37 we get

$$[\mathcal{L}'] \cap [-T] = \partial[\mathcal{L}'] \cap \partial[-T] = [\mathcal{U}(-T, T)] = \partial[-T] .$$

Thus (ii) will follow, once we have shown that $\partial(-T] \supseteq \partial[\mathcal{L}']$. But this is easy: If $[Z] \subseteq \partial[\mathcal{L}']$, let $[Y] \supseteq [Z]$ be any cell of dimension $n-1$. By Theorem 8.13, $[Y]$ is contained in precisely two topes $[T']$ and $[T'']$. One of these must be equal to $[-T]$, since otherwise both $[T']$ and $[T'']$ were in $\mathcal{L}'$, implying that $[Y]$, and hence $[Z] \notin \partial[\mathcal{L}']$. Hence $[Y]$ is a facet of $[-T]$, i.e. $[Z] \subseteq [Y] \subseteq \partial[-T]$.

(iii) is an immediate consequence of (ii) and the definition of B-shellability.

□

Finally, let us mention one more constructibility result, mainly in order to prepare our discussion of the relationship between oriented matroids and "sphere systems" as introduced in chapter 9.

Definition 8.39 *Let $\mathcal{O}$ be a (simple) OM on E. For $e \in E$, define*

$$\begin{aligned} H_e^0 &:= \{X \in \mathcal{O} \mid X_e = 0\}, \\ H_e^+ &:= \{X \in \mathcal{O} \mid X_e = +\} \quad \textit{and} \\ H_e^- &:= \{X \in \mathcal{O} \mid X_e = -\}. \end{aligned}$$

Furthermore, define $\bar{H}_e^+ := H_e^+ \cup H_e^0$ and $\bar{H}_e^- := H_e^- \cup H_e^0$. It is easy to see that $\bar{H}_e^+ = [H_e^+]$, the order ideal generated by H_e^+ and similarly $\bar{H}_e^- = [H_e^-]$. The sets $\bar{H}_e^+$ and $\bar{H}_e^-$ are called **closed halfspheres** *of $\mathcal{O}$. The sets H_e^0 are called* **hyperspheres**. *The intersection of a set of hyperspheres is called a* **flat**. *The intersection of a set of closed halfspheres, provided it is not a flat, is called a* **supercell**. *If K is a supercell, then*

$$F := \bigcap \{H_e^0 \mid K \subseteq H_e^0\}$$

is called the flat **generated** *by K. The* **dimension** *of a flat F or a supercell K is defined to be the maximal dimension of one of its cells. This is denoted by* dim F *resp.* dim K. *Figure 8.6 below shows a 2-dimensional supercell made up of three 2-dimensional cells.*

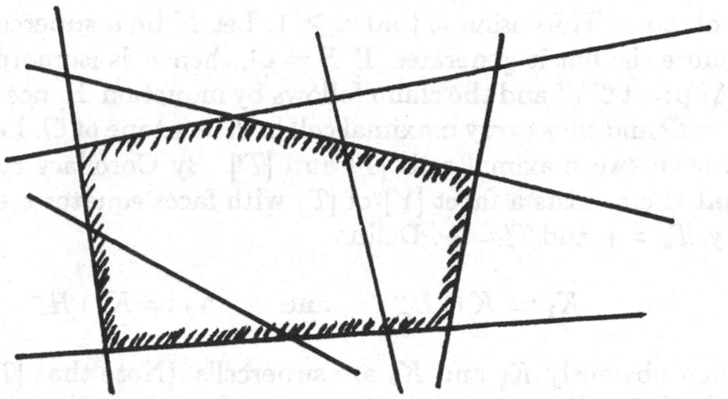

Figure 8.6

It is easy to see that all maximal elements in a flat or a supercell have the same dimension. If F is a flat, say,

$$F = \bigcap\{H_e^0 \mid e \in I\} \qquad \text{for some } I \subseteq E\,,$$

then F, as a suborder of $\mathcal{O}$, is isomorphic to $\mathcal{O}\setminus I$. Hence, in particular, all maximal cells in F have dimension equal to $\dim F = \dim(\mathcal{O}\setminus I)$. Now let K be a supercell, say

$$K = \bigcap\{H_e^+ \mid e \in A\} \qquad \text{for some } A \subseteq E\,,$$

then define

$$I := \{e \in E \mid K \subseteq H_e^0\}\,.$$

By definition, $F = \bigcap\{H_e^0 \mid e \in I\}$ is the flat generated by K. If $[X]$ is any maximal cell in K, then $[X]$ is a tope in $F \cong \mathcal{O}\setminus I$. In fact, suppose $X_e = 0$ for some $e \notin I$. Then there exists a $Y \in K$ such that $Y \notin H_e^0$. Since K is closed under composition, we get $X \circ Y \in K$, contradicting the maximality of X. Thus we have shown that indeed every maximal cell in K is a tope in $F = \mathcal{O}\setminus I$ and hence its dimension equals $\dim F$.

Since every flat F is isomorphic to some OM poset $\mathcal{O}\setminus I$, we know that F is S-constructible. In the following we will show that every supercell is B-constructible, thereby generalizing B-constructibility of (single) cells. The inductive step in the proof is provided by

Lemma 8.40 *Let K be a supercell. Then either K is a cell $[X]$ of $\mathcal{O}$ or there exist two supercells K_1 and K_2 such that $K = K_1 \cup K_2$ and $K_1 \cap K_2 = \partial K_1 \cap \partial K_2$ is a supercell of dimension* $\dim K - 1$.

Proof. The proof is by induction on $n := \dim \mathcal{O}$. If $n = 0$, the claim is obvious. Thus assume that $n \geq 1$. Let K be a supercell and let F denote the flat it generates. If $F \neq \mathcal{O}$, then F is isomorphic to some OM poset $\mathcal{O}\backslash I$ and the claim follows by induction. Hence assume that $F = \mathcal{O}$ and thus every maximal cell in K is a tope of $\mathcal{O}$. Let K contain at least two maximal cells $[T]$ and $[T']$. By Corollary 8.17 we know that there exists a facet $[Y]$ of $[T]$ with facet equator $e \in sep(T, T')$, say, $T_e = +$ and $T'_e = -$. Define

$$K_1 := K \cap \bar{H}_e^+ \qquad \text{and} \qquad K_2 := K \cap \bar{H}_e^- .$$

Then obviously K_1 and K_2 are supercells. (Note that $[T] \subseteq K_1$ and $[T'] \subseteq K_2$. Hence if, say, K_1 were a flat, then $K_1 = \mathcal{O}$, which is ridiculous.) Furthermore, we have $K = K_1 \cup K_2$. Similarly, $K_1 \cap K_2$ is easily seen to be a supercell. Since $[Y] \subseteq K_1 \cap K_2 \subseteq H_e^0$, its dimension equals $n - 1 = \dim K - 1$. This also shows that the flat generated by $K_1 \cap K_2$ is H_e^0. Thus, if $[X]$ is any maximal cell in $K_1 \cap K_2$, then $[X]$ is a tope in $\mathcal{O} \setminus e$, i.e. $supp\, X = E \setminus e$. From this it is clear that from the two unique elements covering X, one is in K_1 and the other is in K_2. This means that $[X] \subseteq \partial K_1 \cap \partial K_2$. Since this holds for every maximal cell in $K_1 \cap K_2$, we get $K_1 \cap K_2 = \partial K_1 \cap \partial K_2$.

□

Proposition 8.41 *Every supercell K is B-constructible of dimension* $\dim K$. *Hence, in particular, every closed halfsphere $\bar{H}_e^+$ is B-constructible of dimension $n = \dim \mathcal{O}$.*

Proof. The obvious inductive argument applies. If a supercell consists of a single maximal cell $[X]$, then it is isomorphic to some tope lattice and hence it is B-constructible by Theorem 8.35. If K contains at least two maximal cells, the claim follows from Lemma 8.40 and the definition of B-constructibility.

□

A simple corollary of the above constructibility results is the following:

Definition 8.42 *Let P be a JD poset of dimension $d = \max\{\dim(p) \mid p \in P\}$. Let f_i $(0 \leq i \leq d)$ denote the number of elements of dimension i. Then*

$$\chi(P) := \sum_{i=0}^{d} (-1)^i f_i$$

is called the **Euler characteristic** *of P.*

Theorem 8.43 *[*EULER's Formula*] If P is B-constructible, then $\chi(P) = 1$, and if P is S-constructible of dimension d, then $\chi(S) = 1 + (-1)^d$.*

Proof. The proof is by induction on d. The claim is true for $d = -1$. Thus, let $d \geq 0$. The following are obvious:

ad (i): If P has a unique maximal element, then

$$\chi(P) = (-1)^d + \chi(\partial P).$$

ad (ii): If P is obtained by pasting two B-constructible posets P_1 and P_2, then

$$\chi(P) = \chi(P_1) + \chi(P_2) - \chi(P_1 \cap P_2).$$

This proves the claim by induction.

□

In particular, if P is a 3-dimensional polytope and f_0, f_1, f_2 denote the number of faces of dimension $0, 1, 2$, resp. (i.e. vertices, edges and facets), then

$$f_0 - f_1 + f_2 = 1 + (-1)^2 = 2.$$

Thus Theorem 8.43 is a generalization of EULER's Formula for planar graphs (cf. Section 2.4), as can be seen from STEINITZ' Theorem (cf. Section 8.5 below).

8.5 Further Reading

The material covered in Chapter 8 is contained in

A. MANDEL *Topology of Oriented Matroids*, Thesis, University of Waterloo, Canada (1981), supervised by J. Edmonds.

Further properties of oriented matroid face lattices, operations on face lattices, such as join, vertex pulling etc. may be found in the above thesis or in the work by Munson (cf. [135]). In particular, Munson's result about nonexistence of polars (cf. Section 7.9) shows that tope lattices are proper generalizations of polytope face lattices, i.e. there do exist tope lattices which are not isomorphic to any polytope face lattice.

The problem of characterizing face lattices of polytopes (cf. Section 7.5) is one of the most challenging problems in polyhedral theory. The 3-dimensional case can be settled by means of Steinitz' wellknown theorem: Say that a (undirected) graph $G = (V, E)$ is $\mathbb{R}^3$-realizable if there exists a polytope P whose vertices and edges (faces of dimension 0 and 1, resp.) are in 1-1 correspondence with V and E and such that the incidence relation between vertices and edges is preserved. Then Steinitz' Theorem can be stated as follows:

Suppose $G = (V, E)$ is 3-connected, i.e. removing any two vertices of and all edges incident to these leaves a connected graph. Then G is planar if and only if G is $\mathbb{R}^3$-realizable.

For higher dimensional lattices, no straightforward analogue can hold, cf. B. STURMFELS [150]. As we mentioned already, tope lattices are strictly more general structures than face lattices of polyhedral cones. Hence, in particular, all wellknown properties of face lattices of topes (such as relative complementarity, shellability) are not sufficiently strong to characterize polytope face lattices.

Nonetheless, in order to have at least a chance of being a polytope face lattice, a given lattice L of course must be a tope lattice at least. This may help to decide the problem whether a given lattice L can be realized by means of a polytope, cf.

J. BOKOWSKY, B. STURMFELS *Coordinatization of Oriented Matroids*, Discrete and Computational Geometry 1(4) (1986).

It can be shown by means of Tarski's Theorem, that the above realizability problem is decidable, i.e. there is an algorithm which, given $(L, \leq)$ as input, decides in a finite number of steps whether L is realizable as a polytope face lattice or not, cf.

A. TARSKI *A Decision Method for Elementary Algebra and Geometry*, Berkley (1951),

as well as the earlier mentioned book by Grünbaum. The purpose of the above mentioned work of Bokowski and Sturmfels is to develop an algorithm which solves the realizability problem at least for small lattices $(L, \leq)$ in a reasonable amount of time (which Tarski's "General Purpose" Algorithm does not).

Chapter 9
Topological Realizations

Let $B \in \mathbf{R}^{E \times n}$ and let $\mathcal{O} = \sigma(im\, B)$ denote the corresponding OM. Suppose for simplicity that B has no zero rows, thus every $H_e^0 = \{x \in \mathbf{R}^n \mid B_e.x = 0\}$ is a hyperplane in $\mathbf{R}^n$. This family of hyperplanes $(H_e \mid e \in E)$ subdivides $\mathbf{R}^n$ into the family of polyhedral cones definable from $Bx \leq 0$. In fact, as we have seen in Section 7.5, the canonical map $\sigma \circ B : \mathbf{R}^n \to \mathcal{O}$ induces a 1-1 correspondence between the cells $[Y] \subseteq \mathcal{O}$ and the corresponding cones $C_Y \subseteq \mathbf{R}^n$ such that $X \preceq Y$ in $\mathcal{O}$ iff $C_X \leq C_Y$, i.e. C_X is a face of C_Y. Thus in case $\mathcal{O}$ is a linear OM, i.e. $\mathcal{O} = \sigma(im\, B)$ for some matrix B, the purely combinatorial structure $\mathcal{O}$ can be "realized" by a topological structure consisting of a family $(C_Y \mid Y \in \mathcal{O})$ of polyhedral cones as above. The purpose of this chapter is to present a similar kind of topological realization for general OMs.

Section 9.1 revisites the canonical map $\sigma \circ B$ and slightly modifies the above "realization", introducing the notion of "linear sphere systems" in order to prepare the more general concepts to be defined in the sequel. Section 9.2 presents an example of a nonlinear OM, i.e. one which cannot be written as $\mathcal{O} = \sigma(im\, B)$ for any matrix B. Section 9.3 introduces the concept of (general) sphere systems as a topological analogue of the linear sphere systems of Section 9.1.

9.1 Linear Sphere Systems

Let $B \in \mathbf{R}^{E \times n}$ and let $\mathcal{O} = \sigma(im\, B)$ the correspoding OM on E. Recall from Section 7.5 that the canonical map $\sigma \circ B : \mathbf{R}^n \to \mathcal{O}$ gives a 1-1 correspondence between cells $[Y] \subseteq \mathcal{O}$ and polyhedral cones in $\mathbf{R}^n$, definable from $Bx \leq 0$. For $Y \in \mathcal{O}$ let $C_Y \subseteq \mathbf{R}^n$ denote the cell (i.e. the polyhedral cone) corresponding to $[Y]$, i.e. $C_Y = (\sigma \circ B)^{-1}[Y]$.

We have indicated already in Section 7.5 that the "combinatorial structure" of $\mathcal{O}$ is reflected in the geometrical (or topological) structure of the system $\mathcal{S} = (C_Y \mid Y \in \mathcal{O})$ in some sense. In particular, the "combinatorial" relation $\preceq$ in $\mathcal{O}$ corresponds to the "topological" relation $\leq$ (i.e. "is a face of") in $\mathcal{S}$. In this section we will analyze the topological structure of the system $\mathcal{S}$ (and a related system to be defined later on). The results are not very surprising and we state them only in order to prepare the definition of a more general class of "cell systems" in Section 9.4.

Proposition 9.1 *Let $\mathcal{S} = (C_Y \mid Y \in \mathcal{O})$, and for each $Y \in \mathcal{O}$ let $s(Y) := \operatorname{int} C_Y$. Then the following holds:*

(i) *$s(0)$ is a subspace of $\mathbb{R}^n$.*

(ii) $\mathbb{R}^n = \dot{\bigcup}\{s(Y) \mid Y \in \mathcal{O}\}$.

(iii) *For every $Y \neq 0$ the set $C_Y = \bigcup\{s(X) \mid X \in [Y]\}$ is a polyhedral cone with $\partial C_Y = \bigcup\{s(X) \mid X \in \partial[Y]\}$.*

Proof. (i) is trivial, and (ii) and (iii) are proved in Theorem 7.32. □

One can show that every polyhedral cone $C \subseteq \mathbf{R}^n$ –provided $C \neq lin(C)$– is homeomorphic to a closed halfspace $\overline{H^+} \subseteq \mathbf{R}^d$ where $d = \dim C$. Thus, from a topological point of view, Proposition 9.1 provides a relationship between $\mathcal{O}$ and a system of "topological halfspaces" satisfying conditions (i)–(iii) above. In the following we are going to replace the system $(C_Y \mid Y \neq 0)$ of topological halfspaces by an "equivalent" system of topological balls. This is not a very essential step, in fact, it is more or less only due to tradition and the fact that balls are, maybe, somewhat handier than halfspaces.

Definition 9.2 *A* **topological *n*-ball**, *or, for short, a* **topological ball** *is a topological space B, homeomorphic to the n-dimensional unit ball $B_n = \{x \in \mathbb{R}^n \mid \|x\| \leq 1\}$. The number n is called the* **dimension** *of B. Similarily, a* **topological *n*-sphere**, *or, for short, a* **topological sphere** *is a topological space S, homeomorphic to the n-dimensional sphere $S^n = \{x \in \mathbb{R}^{n+1} \mid \|x\| = 1\}$. The number n is called the* **dimension** *of S. By convention, $\mathbb{R}^0 = \{0\}$, thus $S^{-1} = \emptyset$ and $B_0 = \{0\}$.*

If B is a topological ball and $h : B_n \to B$ is any homeomorphism then the **boundary** *of B is defined to be*

$$\partial B := h(\partial B_n) = h(S^{n-1})$$

and its **interior** *is defined to be $\operatorname{int} B := B \setminus \partial B$. The* **boundary** *of a topological sphere S is defined to be $\partial S := \emptyset$ and its* **interior** *is $\operatorname{int} S := S$.*

Remark 9.3 Note that the boundary ∂B of a topological ball B is in fact well-defined: If $h : B_n \to B$ and $g : B_m \to B$ are homeomorphisms then $g^{-1}h : B_n \to B_m$ is a homeomorphism. From this one concludes that $m = n$ and $g(\partial B_m) = h(\partial B_m)$ by means of the Domain Invariance Theorem (cf. Section 1.3).

Lemma 9.4 *Let $C \subseteq \mathbb{R}^n$ be a polyhedral cone. Then $C \cap S^{n-1}$ is a topological sphere or a topological ball, according to whether or not C is a subspace of $\mathbb{R}^n$.*

Proof. Elementary proofs of this result are straightforward but somewhat tedious. We leave it as an exercise. Shorter proofs can be obtained by applying the PL-techniques which we are going to sketch in Section 9.4.

□

Now we are ready to replace our original system $(C_Y \mid Y \in \mathcal{O})$ by the system $(C_Y \cap S^{n-1} \mid Y \in \mathcal{O})$. Instead of considering the partition of $\mathbf{R}^n$ induced by the system of hyperplanes $H_e^0 = \{x \in \mathbf{R}^n \mid B_{e.}x = 0\}$, we consider the partition of S^{n-1}, induced by the system of "hyperspheres" $H_e^0 \cap S^{n-1}, e \in E$.

Definition 9.5 *Let $L \leq \mathbb{R}^n$ be a subspace. Then $S := L \cap S^{n-1}$ is called a* **linear sphere**. *If $S = L \cap S^{n-1}$ is a linear sphere and $c \in \mathbb{R}^n \setminus \{0\}$ such that $H^0 := \{x \in S \mid cx = 0\} \neq S$ then H^0 is called a* **linear hypersphere** *of S. $H^+ := \{x \in S \mid cx > 0\}$ and $H^- := \{x \in S \mid cx < 0\}$ are called the* **sides** *of H^0 in S. The triple (H^0, H^+, H^-) is then called an* **oriented linear hypersphere** *of S. A* **linear sphere system** *is a family $(H_e^0, H_e^+, H_e^-)_{e\in E}$ of oriented linear hyperspheres of S^{n-1}. The sets $\overline{H_e^+} := H_e^+ \cup H_e^0$ and $\overline{H_e^-} := H_e^- \cup H_e^0$ are called* **closed halfspheres**. *If $A \subseteq E$ then $F = \bigcap\{H_e^0 \mid e \in A\}$ is called a* **flat** *of the system. The intersection of any set of closed halfspheres which is not a flat is called a* **supercell**. *A supercell which is not a proper union of two or more other supercells is called a* **cell**.

From Proposition 9.1 and Lemma 9.4 we get immediately:

Theorem 9.6 *Let $(H_e^0, H_e^+, H_e^-)_{e\in E}$ be a linear sphere system. Then the following holds:*

(i) *Any flat is a linear sphere.*

(ii) *If F is a flat and $F \not\subseteq H_e^0$ then $H_e^0 \cap F$ is a linear hypersphere of F with sides $H_e^+ \cap F$ and $H_e^- \cap F$.*

(iii) *Any supercell is a topological ball.*

Proof. (i) and (ii) are trivial, and (iii) is the content of Lemma 9.4. □

Of course, almost nothing has happend so far. Essentially, all we did was to replace the hyperplanes $\{x \in \mathbb{R}^n \mid B_e.x = 0\}$ by the linear hyperspheres $\{x \in S^{n-1} \mid B_e.x = 0\}$ and the cones C_Y by their corresponding cells $C_Y \cap S^{n-1}$. But, as we noted already, the main purpose of our work was to prepare the introduction of (general) sphere systems in Section 9.3.

9.2 A Nonlinear OM

A natural question that comes up as soon as oriented matroids are defined, is whether we really got something more general than just systems of sign vectors arising from linear subspaces. Now the time has come to answer this question: There do exist oriented matroids that can not be written as $\mathcal{O} = \sigma(im\,B)$ with $B \in \mathbb{R}^{E \times n}$. We do not intend to give a rigorous proof of this (which would be rather lengthy), but just sketch an example of one such oriented matroid.

Consider a system of ten lines in the plane $\mathbb{R}^2$ as indicated below:

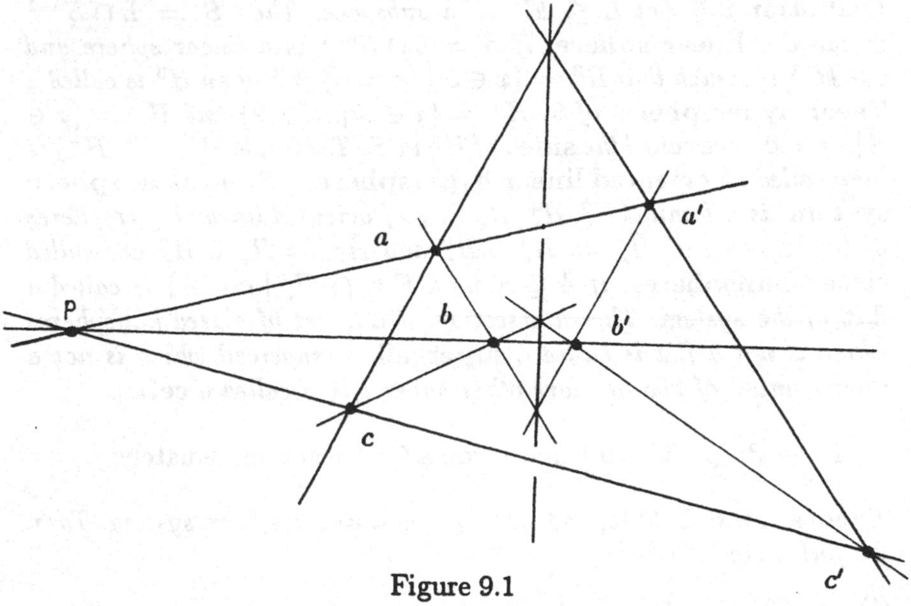

Figure 9.1

This is called "DESARGUES Configuration" and a wellknown theorem from geometry ("DESARGUES Theorem") says that in any such configuration the three lines aa', bb' and cc' meet in a point p. Now

imagine this system of lines (with the three lines meeting in p) being embedded into the 2-dimensional sphere $S^2 \subseteq \mathbb{R}^3$ (with the lines replaced by 1-dimensional linear spheres), thus giving rise to a linear sphere system as introduced in Section 9.1. The corresponding oriented matroid is a linear one, which can be written as $\mathcal{O} = \sigma(im\, B)$ for some $B \in \mathbb{R}^{10\times3}$. Recall that every $Y \in \mathcal{O}$ corresponds to a cell in the linear sphere system (i.e. Y represents a "possible" position of a point on S^2 with respect to the system of hyperspheres).

Now we slightly deform one of the three lines meeting at p, thereby making it pass by the point p. This gives rise to a new system of "lines" on S^2 which looks like indicated below:

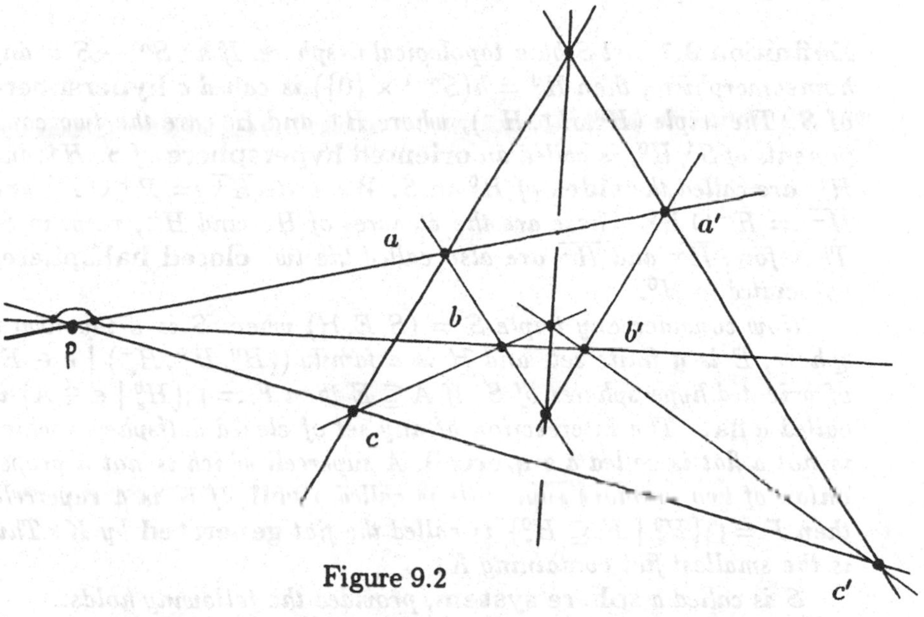

Figure 9.2

Of course, the corresponding operation has to be performed at the antipodal point $-p$ as well. This gives rise to a new system of "cells", i.e. possible positions of a point on S^2 with respect to the modified system of lines. Thus, it is intuitively clear, how to derive a system of sign vectors from this, each sign vector representing a possible position of a point on S^2 with respect to the modified system of lines. One can show by brute force that the system of sign vectors arising that way is an oriented matroid $\mathcal{O}$. We won't do so, however, since we are going to prove a more general result in Section 9.3, which says that any such system of topological lines (or more generally "hyperspheres") on a sphere gives rise to an oriented matroid.

The above oriented matroid $\mathcal{O}$ is called the "Non-DESARGUES Matroid". This cannot be written as $\mathcal{O} = \sigma(im\, B)$ for any matrix B.

In fact, if $\mathcal{O}$ were linear, this would imply that $\mathcal{O}$ can be represented by a linear sphere system, contradicting DESARGUES Theorem.

9.3 Sphere Systems

As we have indicated in Section 9.2, an OM can in general not be "realized" by a linear sphere system. In this section we will introduce a more general class of sphere systems which will turn out to be large enough to allow realizations of general (nonlinear) OMs. This generalization essentially consists in replacing the linear spheres by topological ones.

Definition 9.7 *Let S be a topological n-sphere. If $h : S^n \to S$ is any homeomorphism, then $H^0 = h(S^{n-1} \times \{0\})$ is called a* **hypersphere** *of S. The triple (H^0, H^+, H^-), where H^+ and H^- are the two components of $S \setminus H^0$, is called an* **oriented hypersphere** *of S. H^+ and H^- are called the* **sides** *of H^0 in S. We write $\overline{H^+} := H^+ \cup H^0$ and $\overline{H^-} := H^- \cup H^0$. These are the closures of H^+ and H^-, resp. in S. Therefore, $\overline{H^+}$ and $\overline{H^-}$ are also called the two* **closed halfspheres** *associated to H^0.*

Now consider any triple $\mathcal{S} = (S, E, \mathcal{H})$ where S is a topological sphere, E is a finite set, and $\mathcal{H}$ is a family $((H_e^0, H_e^+, H_e^-) \mid e \in E)$ of oriented hyperspheres of S. If $A \subseteq E$ then $F := \bigcap\{H_e^0 \mid e \in A\}$ is called a **flat**. *The intersection of any set of closed halfspheres which is not a flat is called a* **supercell**. *A supercell which is not a proper union of two or more supercells is called a* **cell**. *If K is a supercell, then $F = \bigcap\{H_e^0 \mid K \subseteq H_e^0\}$ is called the flat* **generated** *by K. This is the smallest flat containing K.*

$\mathcal{S}$ is called a **sphere system**, *provided the following holds:*

(i) *Any flat is a topological sphere.*

(ii) *If F is a flat and $F \not\subseteq H_e^0$ then $F \cap H_e^0$ is a hypersphere of F with sides $F \cap H_e^+$ and $F \cap H_e^-$.*

(iii) *Any supercell is a topological ball.*

Theorem 9.6 states that every linear sphere system is a sphere system. Furthermore, it is intuitively clear that the lines we used to describe the nonlinear OM of Section 9.2 give rise to a sphere system. We will not analyze this here in detail, because we are going to prove in Section 9.4, that, in general, every OM can be "realized" by a sphere system. In this section we will only show that, conversely, every sphere system "realizes" an OM. (A precise definition is straightforward, but will be given below.) Before we can proceed, however, we will have to investigate the topological structure of sphere systems.

Lemma 9.8 *Let $\mathcal{S} := (S, E, \mathcal{H})$ be a sphere system and let $f \in E$. Define*

$$\begin{aligned} \mathcal{H}/f &:= \{(H_e^0, H_e^+, H_e^-) \mid e \in E \setminus f\} \text{ and} \\ \mathcal{S}/f &:= (S, E \setminus f, \mathcal{H}/f) \,. \end{aligned}$$

Then $\mathcal{S}/f$ is a sphere system again. Furthermore, one can define a system $\mathcal{S} \setminus f$ in the obvious way as follows:
Let $\tilde{S} := H_f^0$ (which is a topological sphere) and let

$$\tilde{E} := \{e \in E \mid H_e^0 \neq \tilde{S}\} \,.$$

By axiom (ii) for sphere systems every

$$\tilde{H}_e^0 := H_e^0 \cap \tilde{S} \,, \qquad e \in \tilde{E}$$

is a hypersphere of $\tilde{S}$ with sides

$$\tilde{H}_e^+ := H_e^+ \cap \tilde{S} \quad \text{and} \quad \tilde{H}_e^- := H_e^- \cap \tilde{S} \,.$$

Let

$$\begin{aligned} \mathcal{H} \setminus f &:= ((\tilde{H}_e^0, \tilde{H}_e^+, \tilde{H}_e^-) \mid e \in \tilde{E}) \text{ and} \\ \mathcal{S} \setminus f &:= (\tilde{S}, \tilde{E}, \mathcal{H} \setminus f) \,. \end{aligned}$$

Then $\mathcal{S} \setminus f$ is a sphere system again.

Proof. This is left to the reader as an easy exercise.

□

Remark Be sure that you have well understood the meaning of Definition 9.7. In particular, (iii) states that every supercell K, considered as a topological subspace of S, is homeomorphic to some d-dimensional ball $B^d \subseteq \mathbf{R}^d$. The topology of K is that induced by S, which is the same as that induced by F, where F is the flat generated by K. (This is due to transitivity of induced topologies, cf. Section 1.3).

Since every supercell K is a topological ball, its interior *int* K is defined according to Definition 9.2. For the moment, however, let us consider $int_F K$, the interior of K within the flat F it generates. (This will turn out to be equal to *int* K in a minute.)

Lemma 9.9 *Let K be any supercell, say,*

$$K = \bigcap\{\overline{H_e^+} \mid e \in A\}$$

for some subset $A \subseteq E$ and let F denote the flat generated by K. Then

$$int_F K = \bigcap\{H_e^+ \mid e \in A,\ K \not\subseteq H_e^0\} \cap F\ .$$

Proof. Let $e \in A$ such that $K \not\subseteq H_e^0$. Hence, in particular, $F \not\subseteq H_e^0$. Thus $H_e^+ \cap F$ is open in F. This shows that

$$U := \bigcap\{H_e^+ \mid e \in A,\ K \not\subseteq H_e^0\} \cap F$$

is open in F and contained in K. Hence $U \subseteq int_F K$ (which, by definition, is the largest F-open subset of K). On the other hand, if V is open in F and $V \subseteq K$, then, in particular $V \subseteq F \cap \overline{H_e^+}$ for every $e \in A$. Since V is open in F, we further conclude that

$$V \subseteq int_F(F \cap \overline{H_e^+}) = F \cap H_e^+ \quad \text{for every } e \in A\ .$$

Hence

$$V \subseteq \bigcap\{H_e^+ \mid e \in A\} \cap F \subseteq U\ .$$

In particular, this holds for $V := int_F K$. Hence $U = int_F K$, q.e.d..

□

Proposition 9.10 *Let K be a supercell, say, $K = \bigcap\{\overline{H_e^+},\ e \in A\}$ for some $A \subseteq E$, and let F denote the flat it generates. Then the following hold:*

(i) *$int_F K \neq \emptyset$*

(ii) *$int_F K$ is connected, i.e. it cannot be partitioned into two disjoint F-open sets.*

(iii) *$cl_F(int_F K) = K$*

(iv) $\dim K = \dim F$

(v) *$int_F K = int\, K$*

(vi) *$F = \bigcap\{H_e^0 \mid e \in A,\ K \subseteq H_e^0\}$* .

Proof. We will first prove (i)–(v). Write

$$K = \bigcap\{\overline{H_e^+} \mid e \in A\} \cap F\ .$$

W.l.o.g. we may assume that $K \not\subseteq H_e^0$ for every $e \in A$. The proof of (i)–(v) will be by induction on $k := |A|$. For $k = 1$, i.e. $A = \{e\}$, the claims (i)–(v) follow immediately from the fact that $H_e^0 \cap F$ is a hypersphere with sides $H_e^+ \cap F$ and $H_e^- \cap F$. Assume now that $k \geq 2$. Let $a \in A$ and consider

$$M := \bigcap\{\overline{H_e^+} \mid e \in A \setminus a\} \cap F \,.$$

Thus $K \subseteq M \subseteq F$. If M were a flat, then $M = F$ and $K = \overline{H_a^+} \cap F$, contradicting our assumption that $|A| \geq 2$. Hence M is a supercell and the flat generated by M is F. Furthermore, since $M \supseteq K$, we have that $M \not\subseteq H_e^0$ for every $e \in A \setminus a$. Hence, by induction, (i)–(v) hold for M. From this we will conclude that (i)–(v) also holds for K.

ad (i): By Lemma 9.9, $int_F K = int_F M \cap H_a^+$. Assume that $int_F K = \emptyset$, i.e. $int_F M \subseteq \overline{H_a^-}$. Since the induction applies to M, we get $M = cl_F(int_F M) \subseteq cl_F \overline{H_a^-} = \overline{H_a^-}$. But then $K \subseteq M \cap \overline{H_a^+} \subseteq H_a^0$, contradicting our assumption. Thus (i) holds for K.

ad (ii)–(v): Choose $x \in F \setminus K$. Since F is a topological sphere, there exists a homeomorphism $h : F \to S^m \subseteq \mathbf{R}^{m+1}$. Then $F \setminus x \to S^m \setminus h(x) \simeq \mathbf{R}^m$ (by stereographic projection) gives a homeomorphism $g : F \setminus x \to \mathbf{R}^m$ mapping K onto some subset $B := g(K) \subseteq \mathbf{R}^m$. Now B, being a homeomorphic image of K, is of course a topological ball. Furthermore, we have $int_{\mathbf{R}^m}(B) = g(int_F K) \neq \emptyset$, hence B is in fact m-dimensional. Thus $int\, K = g^{-1}(int_{\mathbf{R}^m} B) = int_F K$ is connected, and its closure in F equals $g^{-1}(B) = K$.

ad (vi): By definition, the flat generated by K equals

$$F = \bigcap\{H_e^0 \mid e \in E,\ K \subseteq H_e^0\} \,.$$

We have to show that this equals

$$\tilde{F} := \bigcap\{H_e^0 \mid e \in A,\ K \subseteq H_e^0\} \,.$$

Obviously, $F \subseteq \tilde{F}$, hence it suffices to show that $\dim F = \dim \tilde{F}$. This will be proved by induction on $k := |E \setminus A|$. If $k = 0$, there is nothing to show. Thus assume $k \geq 1$ and let $f \in E \setminus A$. Consider the sphere system $\mathcal{S}/f$ obtained by removing the hypersphere H_f^0 as described in Lemma 9.8. K is (still) a supercell with respect to $\mathcal{S}/f$ and, by induction, the flat it generates equals $\tilde{F}$. Hence (vi), applied to the sphere system $\mathcal{S}/f$, yields $\dim \tilde{F} = \dim K$. On the other hand, (iv) applied to our original system $\mathcal{S}$ gives $\dim F = \dim K$, q.e.d..

□

Now we are ready to prove that sphere systems give rise to oriented matroids.

Definition 9.11 *Let $\mathcal{S} = (S, E, \mathcal{H})$ be a sphere system. Define $\sigma : S \to 2^{\pm E}$ by*

$$(\sigma(x))_e := \begin{cases} + & \text{if } x \in H_e^+ \\ - & \text{if } x \in H_e^- \\ 0 & \text{if } x \in H_e^0. \end{cases}$$

Let $\mathcal{O}(\mathcal{S}) := \sigma(S) \cup \{0\}$.

Theorem 9.12 *If $\mathcal{S} = (S, E, \mathcal{H})$ is a sphere system then $\mathcal{O}(\mathcal{S})$ is an oriented matroid.*

Proof. Let $\mathcal{O} = \mathcal{O}(\mathcal{S})$.
Before proving that $\mathcal{O}$ satisfies the OM-axioms, let us make the following simple

Observation: *Let K be an intersection of closed halfspaces, say,*

$$K = \bigcap\{\overline{H_e^+} \mid e \in A\}.$$

If $K \cap H_e^+ \neq \emptyset$ for some $e \in A$, then K is a supercell (and not a flat).

Proof. Suppose $K \cap H_e^+ \neq \emptyset$ but K is a flat. By axiom (ii) for sphere systems, $K \cap H_e^0$ is a hypersphere of K with sides $K \cap H_e^+$ and $K \cap H_e^-$. Thus, in particular $K \cap H_e^-$ is nonempty. But this is impossible, since $K \subseteq \overline{H_e^+}$. ◇

Now let us verify the OM-axioms for $\mathcal{O}$.

(i) $X, Y \in \mathcal{O}$ implies $X \circ Y \in \mathcal{O}$: Let $X, Y \in \mathcal{O}$ with $supp(X) \not\supseteq supp(Y)$ and let $x, y \in S$ such that $X = \sigma(x)$ and $Y = \sigma(y)$. Let

$$K := \bigcap\{\overline{H_e^+} \mid x, y \in \overline{H_e^+}\} \cap \bigcap\{\overline{H_e^-} \mid x, y \in \overline{H_e^-}\}.$$

Then K is easily seen to be a supercell. In fact, take any $e \in supp\, X$, say $e \in X^+$. Then $x \in K \cap H_e^+$, i.e. $K \cap H_e^+$ is nonempty and hence our above observation implies that K is a supercell. Now let

$$M := \bigcap\{\overline{H_e^+} \mid x \in H_e^+,\ y \in H_e^-\} \cap \bigcap\{\overline{H_e^-} \mid x \in H_e^-,\ y \in H_e^+\}$$

and consider $N := M \cap K$. This is a supercell again. Indeed, if $M = K$, there is nothing to show any more. Otherwise, if $N \subset K$, there exists some e such that, say $x \in H_e^+$ and $y \in H_e^-$. But then $x \in N \cap H_e^+ \neq \emptyset$ and our observation again implies that N is a supercell. By Proposition 9.10, $int\, N \neq \emptyset$. Obviously, any $z \in int\, N$ gives $Z := \sigma(z) = X \circ Y$.

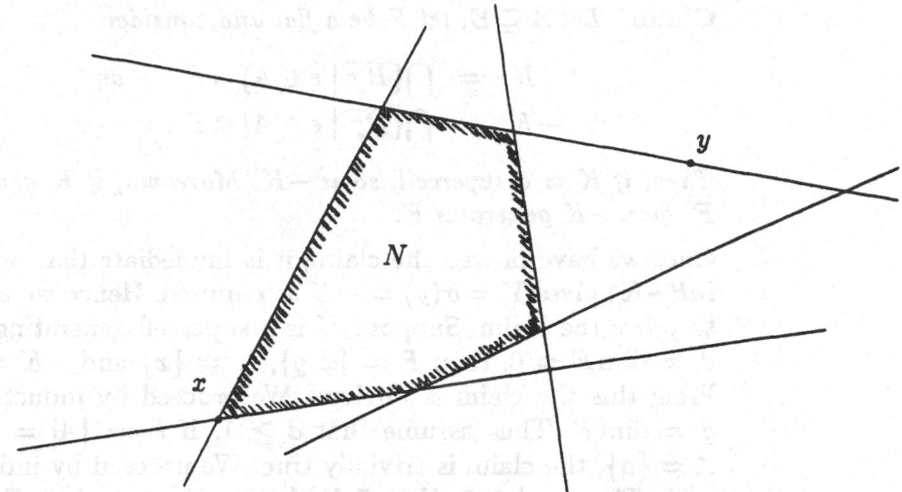

(ii) $\mathcal{O}$ has the approximation property: Let $X, Y \in \mathcal{O}$ and let $x, y \in S$ and K as above. Let $e \in sep(X, Y)$, say $e \in X^+ \cap Y^-$. Let $K_1 := K \cap \overline{H_e^+}$ and $K_2 := K \cap \overline{H_e^-}$. Now $x \in K_1 \cap H_e^+$ and $y \in K_2 \cap H_e^-$. In particular $K_1 \cap H_e^+$ and $K_2 \cap H_e^-$ are both nonempty, implying that K_1 and K_2 are supercells. Hence both have nonempty interior. From Lemma 9.9 we get

$$int\, K_1 = int\, K \cap H_e^+ \quad \text{and} \quad int\, K_2 = int\, K \cap H_e^- .$$

But $int\, K$ is connected (cf. Proposition 0.10) and therefore cannot be the union of two nonempty open subsets $int\, K_1$ and $int\, K_2$. Hence $int\, K \cap H_e^0$ must be nonempty. Now it is easy to see that any $z \in int\, K \cap H_e^0$ gives $Z = \sigma(z)$ as an approximation of X and Y on e. In fact, suppose that $f \in supp\, X \cup supp\, Y$ does not separate X and Y, say $f \in X^+ \setminus Y^-$, i.e. $x \in H_f^+$ and $y \in \overline{H_f^+}$. Thus $K \cap H_f^+$ is nonempty since it contains x. In particular, $K \not\subseteq H_f^0$ and therefore Lemma 9.9 gives that $int\, K \subseteq H_f^+$. This means that $z \in H_f^+$, i.e. $Z_f = +$.

(iii) $X \in \mathcal{O}$ implies $-X \in \mathcal{O}$: Let $0 \neq X \in \mathcal{O}$ and choose $x \in S$ with $\sigma(x) = X$. W.l.o.g. assume that $X \geq 0$. Consider

$$\begin{aligned} K &:= \bigcap\{\overline{H_e^+} \mid x \in H_e^+\} \cap \bigcap\{H_e^0 \mid x \in H_e^0\} \quad \text{and} \\ -K &:= \bigcap\{\overline{H_e^-} \mid x \in H_e^+\} \cap \bigcap\{H_e^0 \mid x \in H_e^0\} . \end{aligned}$$

From our observation, it is clear that K is a supercell.

Claim *Let $A \subseteq E$, let F be a flat and consider*

$$K := \bigcap\{\overline{H_e^+} \mid e \in A\} \cap F \quad \text{and}$$
$$-K := \bigcap\{\overline{H_e^-} \mid e \in A\} \cap F .$$

Then, if K is a supercell, so is $-K$. Moreover, if K generates F, then $-K$ generates F.

Once we have proved the claim, it is immediate that any $y \in int(-K)$ gives $Y = \sigma(y) = -X$ as required. Hence we are left to prove the claim. Suppose K is a supercell generating F. If $d := \dim F = 0$, then $F = \{x, y\}$, $K = \{x\}$ and $-K = \{y\}$. From this the claim is obvious. We proceed by induction on $d = \dim F$. Thus assume that $d \geq 1$. If $k := |A| = 1$, say $A = \{a\}$, the claim is trivially true. We proceed by induction on k. Thus let $k \geq 2$. If $K \subseteq H_a^0$ for some $a \in A$, then $F \subseteq H_a^0$, since K generates F. Hence

$$K = \bigcap\{\overline{H_e^+} \mid e \in A \setminus a\} \cap F \quad \text{and}$$
$$-K = \bigcap\{\overline{H_e^-} \mid e \in A \setminus a\} \cap F ,$$

so the result follows by induction on k in this case. Hence assume that $K \cap H_a^+ \neq \emptyset$ for every $a \in A$. We claim that also $\partial K \cap H_a^+ \neq \emptyset$ for every $a \in A$. Indeed, assume that $\partial K \cap H_a^+ = \emptyset$ for some $a \in A$, i.e. $\partial K \subseteq H_a^0 \cap F$. Pick any $b \in A \setminus a$. Then $K \subseteq \overline{H_b^+}$, hence $\partial K \subseteq \overline{H_b^+} \cap H_a^0 \cap F$, i.e. ∂K is contained in one half of $H_a^0 \cap F$ (we tacitely assumed that the $\overline{H_e^+} \cap F$, $e \in A$, are pairwise distinct). But this is impossible since both ∂K and $H_a^0 \cap F$ are spheres of dimension $d-1$. Hence we get that in fact $\partial K \cap H_a^+ \neq \emptyset$ for every $a \in A$. Now choose any $a \in A$ and $z \in \partial K \cap H_a^+$. Since $z \in \partial K$, we get $z \in H_b^0$ for some $b \in A$. Now consider the sphere system $\mathcal{S} \setminus b$ and let $\tilde{F}$ denote the flat generated by $\tilde{K} := K \cap H_b^0$. Consider

$$\tilde{K} := \bigcap\{\overline{H_e^+} \mid e \in A \setminus b\} \cap \tilde{F}$$
$$-\tilde{K} := \bigcap\{\overline{H_e^-} \mid e \in A \setminus b\} \cap \tilde{F} .$$

Since $z \in \tilde{K} \cap H_a^+$, our observation shows that $\tilde{K}$ is a supercell. Furthermore, since $\tilde{F} \subseteq F \cap H_b^0$, induction on d yields that $-\tilde{K}$ is a supercell generating $\tilde{F}$. In particular, $-\tilde{K} \not\subseteq H_a^0$ (since $\tilde{K} \not\subseteq H_a^0$). This further implies $-K \not\subseteq H_a^0$, i.e. $-K \cap H_a^- \neq \emptyset$. Summarizing, we have proved that $K \cap H_a^+ \neq \emptyset$ implies $(-K) \cap H_a^- \neq \emptyset$. This shows that $-K$ is a supercell generating F.

□

Note the nice relation between the combinatorial properties of $\mathcal{O}$ and the topological properties of sphere systems. In fact, $\mathcal{O}$ is closed under taking sums because $int\, K \neq \emptyset$ for any supercell K. The approximation property corresponds to connectivity of $int\, K$, where K is a supercell and finally, $\mathcal{O} = -\mathcal{O}$ is due to the existence of "antipodal points" in spheres.

9.4 PL Ball Complexes

The main result of Section 9.3 states that any given sphere system $\mathcal{S}$ gives rise to an oriented matroid $\mathcal{O} = \mathcal{O}(\mathcal{S})$ in a natural way. Our final goal in this section is to show that, conversely, given any oriented matroid $\mathcal{O}$, we may construct a sphere system $\mathcal{S}$ such that $\mathcal{O} = \mathcal{O}(\mathcal{S})$.

Definition 9.13 *Consider a triple* $\mathcal{K} = (P, X, s)$, *where* P *is a poset with minimal element* 0, $X \subseteq \mathbb{R}^n$ *for some* n, *and* s *is a map which associates to every* $p \in P$ *a subset* $s(p) \subseteq X$, *called the* **space** *of* p. *If* $Q \subseteq P$, *let* $s(Q) := \bigcup\{s(q) \mid q \in Q\}$. $\mathcal{K}$ *is called a* **topological ball complex** *(or* **ball complex**, *for short) if the following two conditions hold:*

(i) $X = \dot\bigcup\{s(p) \mid p \in P\}$.

(ii) *For every* $p \in P \setminus \{0\}$, $s[p]$ *is a topological ball whose interior is* $s(p)$, *i.e. its boundary is* $s(\partial[p])$. *Hence, in particular,* $s(0)$ *is a topological sphere.*

For $p \in P$ *the set* $s[p]$ *is called a* **cell** *of* $\mathcal{K}$. X *is called the* **space** *of* $\mathcal{K}$, *also denoted by* $s(\mathcal{K})$.

It is easy to see that the intersection of any two cells $s[p]$ and $s[q]$ (with $p \neq q$) can be written as a union of cells in the boundary of $s[p]$ and $s[q]$. In fact, let $x \in s[p] \cap s[q]$ and let $p' \leq p$ be minimal with $x \in s[p']$. Then obviously $x \in s(p')$. Similarly, let $q' \leq q$ be minimal with $x \in s[q']$. Then $x \in s(q')$. Hence from condition (i), we conclude that $p' = q'$ and the claim follows.

Example 9.14 *Let* $P = \{0, p\}$ *consist of just two elements with* $0 < p$. *Let* $X = \{x\}$ *be a single point in* $\mathbb{R}^n$ *and let* $s(0) = \emptyset, s(p) = \{x\}$. *Then* $\mathcal{K} = (P, X, s)$ *is a ball complex. In fact,* $s[p] = \{x\}$ *is homeomorphic to the 0-dimensional ball* $B^0 = \{0\}$ *whose boundary is* $S^{-1} = \emptyset$, *and whose interior is* $int(B^0) = \{0\}$.

Example 9.15 *Let* $P = \{0, p_1, p_2, p\}$ *with a partial order as indicated in the* HASSE *diagram below:*

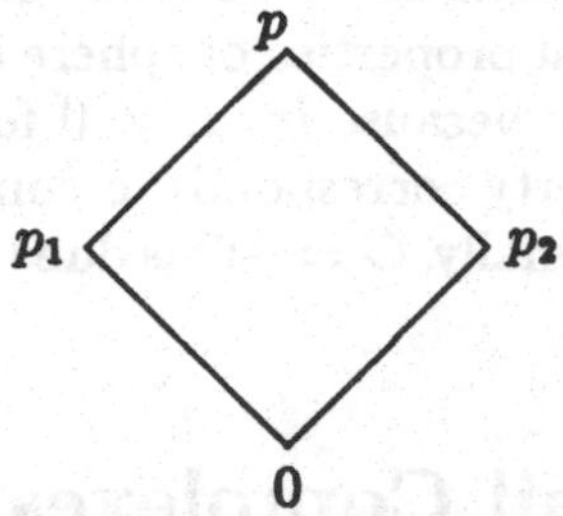

Let X be the line segment $[x_1, x_2]$, joining two points x_1 and x_2 in $\mathbb{R}^n$. Let $s(0) := \emptyset$, $s(p_i) = \{x_i\}$, and $s(p) = (x_1, x_2)$ the open line segment. Then $\mathcal{K} = (P, X, s)$ is a ball complex. In fact, $s[p]$ is a topological ball of dimension 1 and its boundary is $\{x_1, x_2\} = s(p_1) \cup s(p_2)$.

Example 9.16 *Let $P = \{0, p_1, p_2, p, q\}$ with* HASSE *diagram as below:*

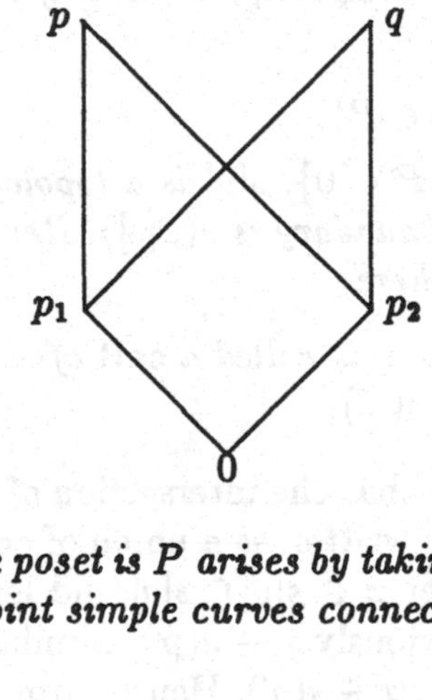

Then a complex whose poset is P arises by taking two points x_1, x_2 in $\mathbb{R}^2$ (say) and two disjoint simple curves connecting x_1 and x_2:

$\mathcal{K} \hat{=}$

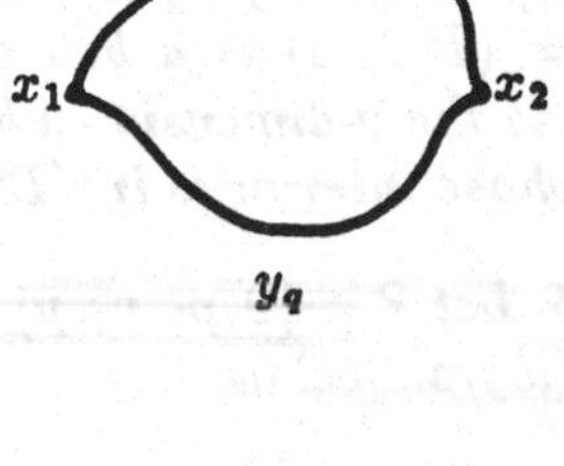

Example 9.17 *Consider a "complex" made up of two 2-dimensional cells and several 0- and 1-dimensional cells as indicated below:*

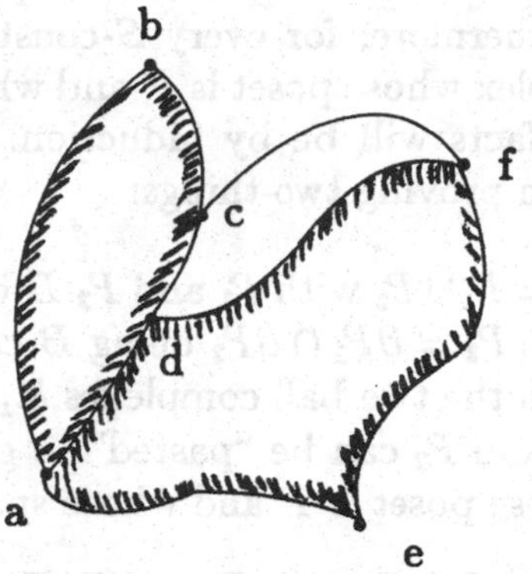

Its poset is given by the HASSE *diagram:*

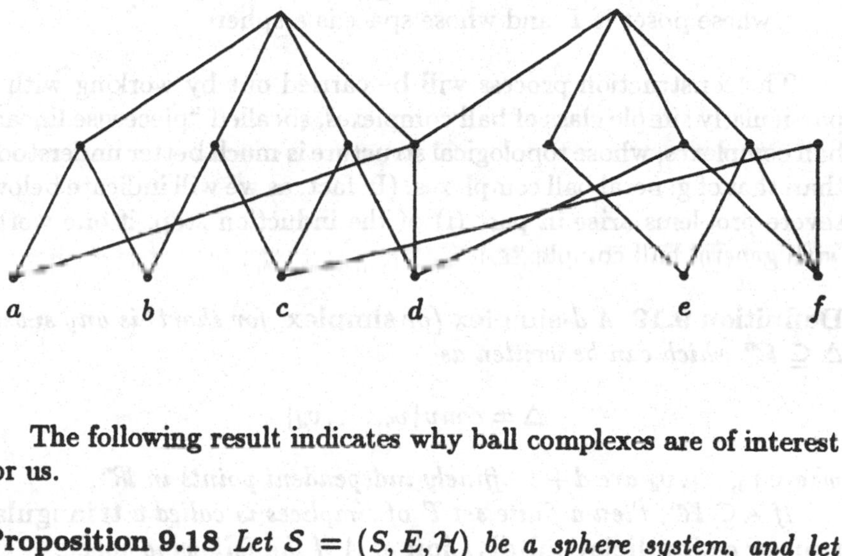

The following result indicates why ball complexes are of interest for us.

Proposition 9.18 *Let $\mathcal{S} = (S, E, \mathcal{H})$ be a sphere system, and let $\sigma : S \to \mathcal{O} \subseteq 2^{\pm E}$ be defined as in Section 9.3. Then $\mathcal{K} = (\mathcal{O}, S, \sigma^{-1})$ is a ball complex.*

Proof. This is an immediate consequence of the results in Section 9.3. In fact, $\sigma^{-1}(0)$ is a (possibly empty) flat of $\mathcal{S}$ and hence it is a sphere. Furthermore, $\sigma^{-1}[Y]$ for every $Y \neq 0$ is a cell whose interior is $\sigma^{-1}(Y)$, as can be seen from Lemma 9.9.

□

Our prime goal in the following will consist in constructing a ball complex $\mathcal{K} = (\mathcal{O}, X, s)$ for a given OM poset $\mathcal{O}$. The basic idea is simple: We will show in fact that for every B-constructible poset P there exists a ball complex whose poset is P and whose space is a ball. Furthermore, for every S-constructible poset P there exists a ball complex whose poset is P and whose space is a sphere. The proof of these facts will be by induction. Essentially, the induction step consists in proving two things:

(i) If $P = P_1 \cup P_2$ with P_1 and P_2 B-constructible of dimension n and $P_1 \cap P_2 = \partial P_1 \cap \partial P_2$ being B-constructible of dimension $n-1$, then the two ball complexes $\mathcal{K}_1$ and $\mathcal{K}_2$ already constructed for P_1 and P_2 can be "pasted" to give a ball complex $\mathcal{K} = \mathcal{K}_1 \cup \mathcal{K}_2$ whose poset is P and whose space is a ball.

(ii) If $P = P_1 \cup P_2$ with P_1 and P_2 B-constructible of dimension n and $P_1 \cap P_2 = \partial P_1 = \partial P_2$ being S-constructible of dimension $n-1$, then the two ball complexes $\mathcal{K}_1$ and $\mathcal{K}_2$ already constructed for P_1 and P_2 can be "pasted" to give a ball complex $\mathcal{K} = \mathcal{K}_1 \cup \mathcal{K}_2$ whose poset is P and whose space is a sphere.

The construction process will be carried out by working with a particularly simple class of ball complexes, socalled "piecewise linear" ball complexes, whose topological structure is much better understood than that of general ball complexes. (In fact, as we will indicate below, severe problems arise in part (i) of the induction step, if one works with general ball complexes.)

Definition 9.19 *A* ***d*-simplex** *(or* **simplex**, *for short) is any subset* $\Delta \subseteq \mathbb{R}^n$ *which can be written as*

$$\Delta = conv\{v_0, \ldots, v_d\}$$

where $v_0, \ldots, v_d$ *are* $d+1$ *affinely independent points in* $\mathbb{R}^n$.

If $A \subseteq \mathbb{R}^n$, *then a finite set* T *of simplices is called a* **triangulation** *or a* **simplicial subdivision** *of* A *if the following holds:*

(i) $A = \bigcup\{\Delta \mid \Delta \in T\}$.

(ii) $\Delta \in T$ *implies* $\Delta' \in T$ *for every face* Δ' *of* Δ.

(iii) *If* $\Delta, \Delta' \in T$ *then* $\Delta \cap \Delta'$ *is a face of both.*

Now let $A \subseteq \mathbb{R}^m$ *and* $B \subseteq \mathbb{R}^n$, *and let* $f : A \to B$ *a bijection. Then* f *is called a* **PL-homeomorphism** *(PLH) if there exists a simplicial subdivision* T *of* A *such that* $f_{|\Delta}$ *is an affine map for every* $\Delta \in T$.

(Recall that $f : X \to Y$ is affine provided the following holds: If $x_1, \ldots, x_k \in X$ and $\lambda \in \mathbb{R}^k_+$ such that $\sum_i \lambda_i = 1$ then $f(\sum_i \lambda_i x_i) = \sum_i \lambda_i f(x_i)$.)

A subset $B \subseteq \mathbb{R}^n$ is called a **PL m-ball** *(or a* **PL-ball**, *for short) if there exists a m-simplex Δ and a PLH $h : \Delta \to B$. Similarily, $S \subseteq \mathbb{R}^n$ is called a* **PL m-sphere** *(or a* **PL-sphere** *, for short) if there exists an $(m+1)$-simplex Δ and a PLH $h : \partial\Delta \to S$.*

Of course, any PL-ball is a topological ball and every PL-sphere is a topological sphere.

Example 9.20 *The following figure indicates a PLH*

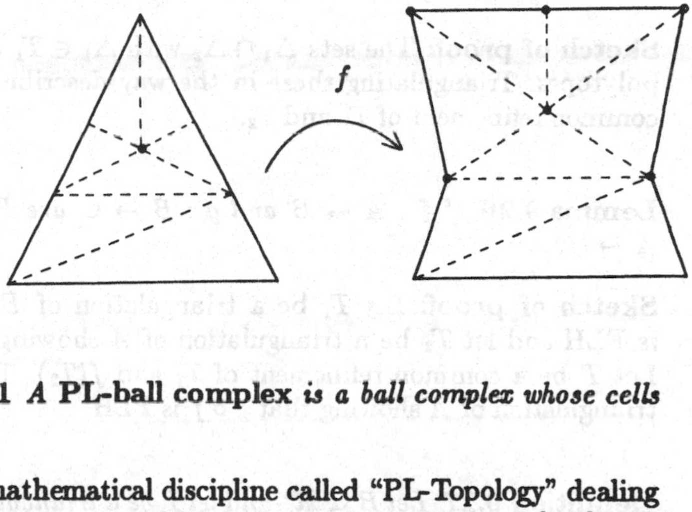

Definition 9.21 *A* **PL-ball complex** *is a ball complex whose cells are PL-balls.*

There is a mathematical discipline called "PL-Topology" dealing with topological properties of PL-ball complexes. HUDSON [103] provides a general introduction into this field. The part which is relevant for our purposes is all contained in Chapter 1 of his book. Therefore, we decided just to present the notions and results we need and sketch the main ideas behind the proofs, referring to HUDSON's book for more details.

Lemma 9.22 *If $f : A \to B$ is a PLH, so is f^{-1}.*

Proof. If T is an appropriate triangulation for f then $f(T)$ is one for f^{-1}.

□

Lemma 9.23 *If $B \subseteq \mathbb{R}^n$ is a polytope, then there exists a triangulation of B.*

Sketch of proof: For any face $\emptyset \neq F$ of B choose $x_F \in int\, F$. Now for any maximal chain $F_0 \lessdot F_1 \lessdot \ldots \lessdot F_k = B$ in the face lattice of B the set $\Delta := conv\{x_{F_0}, \ldots, x_{F_k}\}$ is a simplex and the family T of all these simplices gives a triangulation of B. (This is easily seen by induction on $k = \dim B$.)

□

Definition 9.24 *Let $A \subseteq \mathbb{R}^n$, and let T_1 and T_2 be two triangulations of A. The triangulation T of A is called a* **common refinement** *of T_1 and T_2 if for every $\Delta \in T_1 \cup T_2$ the set $\{\Delta' \in T \mid \Delta' \subseteq \Delta\}$ is a triangulation of Δ.*

Lemma 9.25 *Any two triangulations T_1 and T_2 of a subset $A \subseteq \mathbb{R}^n$ have a common refinement.*

Sketch of proof: The sets $\Delta_1 \cap \Delta_2$ with $\Delta_1 \in T_1$ and $\Delta_2 \in T_2$ are polytopes. Triangulating these in the way described above yields a common refinement of T_1 and T_2.

□

Lemma 9.26 *If $f : A \to B$ and $g : B \to C$ are PLH so is $g \circ f : A \to C$.*

Sketch of proof: Let T_1 be a triangulation of B showing that g is PLH and let T_2 be a triangulation of A showing that f is PLH. Let T be a common refinement of T_1 and $f(T_2)$. Then $f^{-1}(T)$ is a triangulation of A showing that $g \circ f$ is PLH.

□

Definition 9.27 *Let $B \subseteq \mathbb{R}^n$ and let T be a triangulation of B. Then*

$$\partial T := \{\Delta \cap \partial B \mid \Delta \in T\}$$

is called the **triangulation** *of ∂B* **induced** *from T.*

Lemma 9.28 *Let $B \subseteq \mathbb{R}^n$ be a polytope. If T is a triangulation of B then ∂T is a triangulation of ∂B. Conversely, given any triangulation T' of ∂B then there exists a triangulation T of B such that $T' = \partial T$.*

Sketch of proof: The first claim is obvious. To prove the second one, choose $x \in int\, B$ and let

$$T := \{conv(\Delta' \cup x) \mid \Delta' \in T'\}.$$

(Note that B is a polytope, hence a convex set.)

□

Lemma 9.29 *If $B_1 \subseteq \mathbb{R}^m$ and $B_2 \subseteq \mathbb{R}^n$ are polytopes then any PLH $f : \partial B_1 \to \partial B_2$ extends to a PLH $\bar{f} : B_1 \to B_2$.*

Proof. This is obvious from the preceding Lemma 9.28.

□

More generally, the following is true:

Lemma 9.30 *If B_1 and B_2 are PL-balls then any PLH $f : \partial B_1 \to \partial B_2$ extends to a PLH $\bar{f} : B_1 \to B_2$.*

Sketch of proof: Let $h_1 : \Delta^m \to B_1$ and $h_2 : \Delta^n \to B_2$ be PLHs (these exist by definition of PL-balls). Then $f : \partial B_1 \to \partial B_2$ induces a PLH $h_2^{-1} f h_1 : \partial\Delta^m \to \partial\Delta^n$ (implying that $m = n$). By Lemma 9.29, this may be extended to $\bar{h} : \Delta^m \to \Delta^n$. Now $\bar{f} := h_2 \bar{h} h_1^{-1}$ is what we are looking for.

□

Lemma 9.31 *If $\mathcal{K} = (P, X, s)$ is a Pl-ball complex whose space is a Pl-n ball, then*

$$s(\partial P) = \partial X \qquad \textit{(which is a PL-}(n-1)\textit{ sphere).}$$

Sketch of proof: Since X is a PL-n ball, there exists a PLH $f : \Delta^n \to X$. Let T_1 be a corresponding triangulation of X. Construct a second triangulation of X as follows: Initially, let $T_2^{(0)}$ consist of all 0-cells of $\mathcal{K}$. Suppose inductively, that $T_2^{(i)}$, $(i < n)$, is already constructed, inducing a triangulation on every i-cell, i.e. on every cell $s[q]$ which is a PL-i ball. Now if $s[p]$ is a PL-$(i+1)$ ball, then $\partial(s[p]) = s(\partial[p])$ is already triangulated. Using Lemma 9.30, this triangulation can be extended to a triangulation of $s[p]$. Let $T_2^{(i+1)}$ consist of all simplices occuring in such a triangulation of one of the $(i+1)$-cells. Finally, let $T_2 := T_2^{(n)}$. Let T be a common refinement of T_1 and T_2. Thus f is a PLH with respect to T, and T contains a triangulation of every n-cell of $\mathcal{K}$. Since $f : \Delta^n \to X$ is a PLH, it is clear that ∂X consists of precisely those $(n-1)$-simplices in T which are contained in precisely one n-simplex of T. Since T triangulates the cells $s[p]$, $p \in P$, it is easy to see that the boundary of ∂X consists of precisely those $(n-1)$-cells $s[q]$ which are contained in precisely one n-cell $s[p]$, and the result follows.

□

The following is a somewhat more difficult result, also known as "NEWMAN's Theorem" (cf. HUDSON [103]).

Theorem 9.32 *If $S \subseteq \mathbb{R}^n$ is a PL m-sphere and $B \subseteq S$ is a PL m-ball then $cl(S \setminus B)$ is a PL m-ball.*

□

Although this result is intuitively clear, its proof is based on some PL-techniques which we do not want to introduce, since they are of no further use for us. We would like to point out, however, that NEWMAN's Theorem is crucial for our inductive construction of PL-ball complexes representing OM posets as explained below. NEWMAN's Theorem does **not** hold for general topological balls and spheres, (cf. Mandel [126], p.157). The problem is closely related to the socalled "Schoenflies Problem" (cf. Dugundji [64], Chapter XVII Theorem 2.4 ff.): It is known that every topological n-sphere $S \subseteq \mathbf{R}^{n+1}$ separates $\mathbf{R}^{n+1}$. In fact, one can show that $\mathbf{R}^{n+1} \setminus S$ has exactly two components. But it is not true in general that S, together with the bounded component of $\mathbf{R}^{n+1} \setminus S$ is a topological ball! (In fact, the conclusion holds only for $n = 1$.)

Now we have provided enough "PL-tools" in order to study PL-ball complexes and deal with the inductive construction process, showing how to construct large complexes by successively pasting together small ones. For a precise description of the "pasting" procedure, however, we have to introduce some further notations.

Definition 9.33 *Let $\mathcal{K} = (P, X, s)$ be a PL-ball complex, and let $P_1 \subseteq P$ be an order ideal, i.e. $P_1 = [P_1]$. Then P_1 induces a ball complex $\mathcal{K}_1 = (P_1, X_1, s_1)$ whose space X_1 equals $s(P_1)$ in an obvious way. This is called the* **subcomplex** *of $\mathcal{K}$ corresponding to P_1. If $P = P_1 \cup P_2$ for two ideals P_1 and P_2, we also write $\mathcal{K} = \mathcal{K}_1 \cup \mathcal{K}_2$ and say that $\mathcal{K}$ is the* **union** *of $\mathcal{K}_1$ and $\mathcal{K}_2$.* **Intersection** *of two subcomplexes $\mathcal{K}_1$ and $\mathcal{K}_2$ is also defined in the obvious way. The intersection of two subcomplexes $\mathcal{K}_1$ and $\mathcal{K}_2$ is denoted by $\mathcal{K}_1 \cap \mathcal{K}_2$ which is a subcomplex again. The* **boundary** *of $\mathcal{K}$, denoted by $\partial\mathcal{K}$, is the subcomplex associated to ∂P. Two ball complexes are called* **isomorphic**, *provided their posets are isomorphic.*

As we will see in a minute, isomorphic complexes are also essentially the same thing from a topological point of view. In fact, if f is any isomorphism between their posets then there exists a PLH between their spaces which is "compatible" with f in the following sense:

Definition 9.34 *Let $\mathcal{K} = (P, X, s)$ and $\mathcal{K}' = (P', X', s')$ be PL-ball complexes and let $f : p \to p' = f(p)$ be an isomorphism between P*

and P'. Then a PLH $h : s(\mathcal{K}) \to s'(\mathcal{K}')$ is called **compatible** *with f provided $h(s(p)) = s'(p')$ for all $p \in P$, i.e. h takes cells into cells in the way prescribed by f.*

Lemma 9.35 (Isomorphism Lemma) *Let $\mathcal{K} = (P, X, s)$ and $\mathcal{K}' = (P', X', s')$ be PL-ball complexes such that $s(0)$ is PLH to $s'(0')$. Let $f : P \to P'$ be an isomorphism. Then there exists a PLH $h : s(\mathcal{K}) \to s'(\mathcal{K}')$ which is compatible with f. More generally, if $\mathcal{K}_1$ and $\mathcal{K}_1'$ are nonempty subcomplexes of $\mathcal{K}$ and $\mathcal{K}'$, resp., then any PLH between $s(\mathcal{K}_1)$ and $s'(\mathcal{K}_1')$ which is compatible with f can be extended to one between $s(\mathcal{K})$ and $s(\mathcal{K}')$.*

Proof. Let $\mathcal{K}_1$ and $\mathcal{K}_1'$ be subcomplexes and let $h : s(\mathcal{K}_1) \to s'(\mathcal{K}_1')$ be a PLH compatible with f. Let $P_1 \subseteq P$ be the poset of $\mathcal{K}_1$. Let $p \in P \setminus P_1$ be minimal. Then $\partial[p]$ is all contained in P_1. Now h may be extended to the subcomplex corresponding to $P_1 \cup p$ by means of Lemma 9.30. Continuing this way, we end up with an extension $s(\mathcal{K}) \to s(\mathcal{K}')$. (Note that the first step in the induction is provided by our assumption that $s(0)$ is PLH to $s'(0')$.)

□

Theorem 9.36 (Sphere Construction Theorem) *Let P be a poset, and let P_1 and P_2 be two order ideals such that $P = P_1 \cup P_2$ and $P_1 \cap P_2 = \partial P_1 = \partial P_2$. Furthermore, suppose we are given two PL-ball complexes $\mathcal{K}_1$ and $\mathcal{K}_2$ with posets P_1 and P_2, resp., whose spaces are PL n-balls. Then there exists a PL-ball complex $\mathcal{K}$ whose poset is P and whose space is the boundary of a $(n+1)$-simplex (hence a PL n-sphere).*

Proof. Let $e_1, \ldots e_{n+1} \in \mathbf{R}^{n+1}$ denote the $(n+1)$-dimensional unit vectors. Let

$$\begin{aligned} \Delta^n &= conv\{0, e_1, \ldots, e_n\}, \\ \Delta^+ &= conv\{0, e_1, \ldots, e_n, e_{n+1}\} \\ \text{and } \Delta^- &= conv\{0, e_1, \ldots, e_n, -e_{n+1}\}. \end{aligned}$$

Furthermore, let $B_1 := \partial\Delta^+ \setminus int\Delta^n$ and $B_2 := \partial\Delta^- \setminus int\Delta^n$. Obviously, B_1 and B_2 are PL n-balls, and $B_1 \cap B_2 = \partial\Delta^n$ is a PL $(n-1)$-sphere.

Let $\mathcal{K}_1'$ and $\mathcal{K}_2'$ denote the subcomplexes of $\mathcal{K}_1$ and $\mathcal{K}_2$, resp., corresponding to $P_1 \cap P_2 = \partial P_1 = \partial P_2$. Since $s_1(\mathcal{K}_1)$ and $s_2(\mathcal{K}_2)$ are PL n-balls, their boundaries $s_1(\mathcal{K}_1')$ and $s_2(\mathcal{K}_2')$ are PL $(n-1)$-spheres (cf. Lemma 9.31). In particular, there exists a PLH $h' : s_1(\mathcal{K}_1') \to s_2(\mathcal{K}_2')$.

In fact, since the identity map $id : \partial P_1 \to \partial P_2$ is an isomorphism, Lemma 9.35 shows that h' may be choosen to be "cell-preserving", i.e. such that $h'(s_1(p)) = s_2(p)$ for every $p \in P_1 \cap P_2$.

Since $s_1(\mathcal{K}_1')$ and $s_2(\mathcal{K}_2')$ are PL $(n-1)$-spheres there exist PLHs

$$\begin{aligned} h_1 : B_1 \cap B_2 &\to s(\mathcal{K}_1') \text{ and} \\ h_2 := h'h_1 : B_1 \cap B_2 &\to s(\mathcal{K}_2'). \end{aligned}$$

These may be extended to PLHs $\bar{h}_i : B_i \to s_i(\mathcal{K}_i)$. Now we define a PL-ball complex $\mathcal{K}$ whose poset is P and whose space is $B_1 \cup B_2$ in an obvious way. If $p \in P_i$, we define the space of p to be $\bar{h}_i^{-1} s_i(p)$. (If $p \in P_1 \cap P_2$, the two definitions agree.) This gives obviously rise to a PL-ball complex with poset P and space $B_1 \cup B_2$, which is the boundary of a $(n+1)$-simplex.

□

Note that, so far, we have made no use of NEWMANs Theorem. NEWMAN's Theorem will, however, become crucial when we construct PL-balls by pasting balls along part of their boundary, as we are going to do now.

Theorem 9.37 (Ball Construction Theorem) *Let P be a poset and let P_1 and P_2 be order ideals such that $P = P_1 \cup P_2$ and $P_1 \cap P_2 = \partial P_1 \cap \partial P_2$. Furthermore, suppose we are given two PL-ball complexes $\mathcal{K}_1$ and $\mathcal{K}_2$ with posets P_1 and P_2, resp., such that the spaces of $\mathcal{K}_1$ and $\mathcal{K}_2$ are PL n-balls and the spaces of their subcomplexes corresponding to $P_1 \cap P_2$ are PL $(n-1)$-balls. Then there exists a PL-ball complex $\mathcal{K}$ whose poset is P and whose space is an n-simplex (hence a PL n-ball).*

Proof. Let $\mathcal{K}_1'$ and $\mathcal{K}_2'$ be the subcomplexes of $\mathcal{K}_1$ and $\mathcal{K}_2$ corresponding to $P_1 \cap P_2$. Since $s_i(\mathcal{K}_i)$ is a PL n-ball, $s_i(\partial\mathcal{K}_i)$ is a PL $(n-1)$-sphere for $i = 1, 2$ (cf. Lemma 9.31). By assumption, $s_i(\mathcal{K}_i') \subseteq s_i(\partial\mathcal{K}_i)$ is a PL $(n-1)$-ball. Let $\mathcal{K}_i''$ denote the subcomplex of $\partial\mathcal{K}_i$ which corresponds to $[\partial P_i \setminus (P_1 \cap P_2)]$, i.e. the order ideal generated by $\partial P_i \setminus (P_1 \cap P_2)$. Then $s(\mathcal{K}_i'')$ is the closure of $s_i(\partial\mathcal{K}_i) \setminus s_i(\mathcal{K}_i')$, which is a PL $(n-1)$-ball by NEWMAN's Theorem. Furthermore, we must have $\partial\mathcal{K}_i' = \partial\mathcal{K}_i''$, since the space of $\partial\mathcal{K}_i = \mathcal{K}_i' \cup \mathcal{K}_i''$ is a Pl-$(n-1)$ sphere and hence has empty boundary.

Now, since $s_1(\mathcal{K}_1')$ and $s_2(\mathcal{K}_2')$ are PL $(n-1)$-balls, there exists a PLH $f' : s(\mathcal{K}_1') \to s(\mathcal{K}_2')$. By the Isomorphism Lemma, this may be chosen to be cell preserving.

Now let

$$\begin{aligned} \Delta^{n-1} &:= conv\{0, e_1, \dots, e_{n-1}\}, \\ \Delta^+ &:= conv\{0, e_1, \dots, e_{n-1}, e_n\} \quad \text{and} \\ \Delta^- &:= conv\{0, e_1, \dots, e_{n-1}, -e_n\}. \end{aligned}$$

Then there exist PLHs

$$\begin{aligned} h_1' &: \Delta^{n-1} \to s_1(\mathcal{K}_1') \text{ and} \\ h_2' := f' \circ h_1' &: \Delta^{n-1} \to s_2(\mathcal{K}_2') . \end{aligned}$$

These induce PLHs

$$\begin{aligned} g_1 &: \partial\Delta^{n-1} \to \partial s_1(\mathcal{K}_1') = \partial s_1(\mathcal{K}_1'') \\ \text{and} \quad g_2 &: \partial\Delta^{n-1} \to \partial s_2(\mathcal{K}_2') = \partial s_2(\mathcal{K}_2'') \end{aligned}$$

Now let

$$\begin{aligned} B^+ &:= cl((\partial\Delta^+)\setminus\Delta^{n-1}) = (\partial\Delta^+)\setminus int\,\Delta^{n-1} \\ \text{and} \quad B^- &:= cl((\partial\Delta^-)\setminus\Delta^{n-1}) = (\partial\Delta^-)\setminus int\,\Delta^{n-1}. \end{aligned}$$

B^+ and B^- are easily seen to be PL-$(n-1)$ balls. Furthermore, we have $\partial B^+ = \partial B^- = \partial\Delta^{n-1}$. Hence, using Lemma 9.30, g_1 and g_2 can be extended to PLHs

$$\begin{aligned} h_1'' &: B^+ \to s_1(\mathcal{K}_1'') \\ \text{and} \quad h_2'' &: B^- \to s_1(\mathcal{K}_2'') . \end{aligned}$$

Combining h_i' and h_i'' in the obvious way, we get PLHs

$$\begin{aligned} \tilde{h}_1 &: \partial\Delta^+ \to \partial s_1(\mathcal{K}_1) = s_1(\mathcal{K}_1') \cup s_1(\mathcal{K}_1'') \\ \tilde{h}_2 &: \partial\Delta^- \to \partial s_2(\mathcal{K}_2) = s_2(\mathcal{K}_2') \cup s_2(\mathcal{K}_2'') \end{aligned}$$

which agree on Δ^{n-1}. These may again be extended to PLHs

$$\begin{aligned} h_1 &: \Delta^+ \to s_1(\mathcal{K}_1) \quad \text{and} \\ h_2 &: \Delta^- \to s_2(\mathcal{K}_2). \end{aligned}$$

Now define a complex $\mathcal{K}$ whose poset is P and whose space is $\Delta^+ \cup \Delta^-$ (which is an n-simplex) in the obvious way: If $p \in P_i$, define the space of p to be $h_i^{-1}s_i(p)$. If $p \in P_1 \cap P_2$, these definitions agree.

□

The two construction theorems may be put together to yield

Theorem 9.38 (Representation Theorem) *Let P be an n-dimensional constructible poset. If P is B-constructible there exists a PL-ball complex whose poset is P and whose space is Δ^n. If P is S-constructible, there exists a PL-ball complex whose poset is P and whose space is $\partial\Delta^n$.*

Proof. This is immediate from the two construction theorems and induction. Note that if P is B-constructible of dimension n and $P = [p]$ with $\partial[p]$ being S-constructible of dimension $n-1$, then the inductive assumption gives that there exists a PL-ball complex with poset $\partial[p]$ and space $\partial\Delta^n$. In this case, we may simply add p, defining its space to be $int\Delta^n$ to obtain a complex whose poset is P, and whose space is Δ^n.

$\square$

Corollary 9.39 *If $\mathcal{O}$ is the poset of an OM then there exists a PL-ball complex whose poset is $\mathcal{O}$ and whose space is $\partial\Delta^{n+1}$.*

$\square$

From this one easily concludes that OM posets can be represented by sphere systems in the sense of Section 9.3. We present a simple preliminary lemma:

Lemma 9.40 *Let $\mathcal{K} = (P, X, s)$ be a PL-ball complex with $s(0) = \emptyset$. Then the space of a subcomplex of $\mathcal{K}$ corresponding to an n-dimensional B-(S)-constructible order ideal is a PL n-ball (-sphere).*

Proof. Let $\mathcal{K}_1 = (P_1, X_1, s_1)$ be a subcomplex with P_1 being B-constructible of dimension n. By the Representation Theorem there exists a PL-ball complex $\mathcal{K}'_1 = (P_1, \Delta^n, s')$. Hence, by the Isomorphism Lemma there exists a PLH : $\Delta^n \to X_1$ showing that X_1 is a PL n-ball. The case where P_1 is S-constructible is similar.

$\square$

Now let $\mathcal{O} \subseteq 2^{\pm E}$ be an n-dimensional OM which we may assume to be simple. Let $\mathcal{K} = (\mathcal{O}, \partial\Delta^{n+1}, s)$ be a corresponding PL-ball complex. For every $e \in E$ let H_e^0, H_e^+, H_e^-, $\overline{H_e^+}$ and $\overline{H_e^-}$ be defined as in Definition 8.39. We claim that $s(H_e^0)$ is a hypersphere of $s(\mathcal{K}) = \partial\Delta^{n+1}$ with sides $s(H_e^+)$ and $s(H_e^-)$. To see this, recall from Proposition 8.41 that $\overline{H_e^+}$ and $\overline{H_e^-}$ are B-constructible with boundary H_e^0. From Lemma 9.40 we conclude that $s(\overline{H_e^+})$ and $s(\overline{H_e^-})$ are PL-n balls and their boundary (cf. Lemma 9.31) equals $s(H_e^0)$, which is a PL-$(n-1)$-sphere. Now construct a PLH: $\partial\Delta^{n+1} \to \partial\Delta^{n+1}$ showing that $s(H_e^0)$ is a hypersphere as follows. Consider the n-balls

$$B^+ := \Delta^n \subseteq \partial\Delta^{n+1} \qquad \text{and}$$
$$B^- := (\partial\Delta^{n+1}) \setminus (int\,\Delta^n)$$

whose intersection is $B^+ \cap B^- = \partial\Delta^n$. Since $s(H_e^0)$ is an n-sphere, there exists a PLH

$$h^0 : \partial\Delta^n \to s(H_e^0) .$$

This can be extended to PLHs

$$h^+ : B^+ \to s(\overline{H_e^+}) \quad \text{and}$$
$$h^- : B^- \to s(\overline{H_e^-}) \,.$$

Combine h^+ and h^- in the obvious way to a PLH

$$h : \partial\Delta^{n+1} \to s(\mathcal{K}) = \partial\Delta^{n+1},$$

mapping the hypersphere $\partial\Delta^n$ onto the hypersphere $s(H_e^0)$. This shows that in fact $s(H_e^0)$ is a hypersphere of $s(\mathcal{K})$ with sides $s(H_e^+)$ and $s(H_e^-)$.

The remaining axioms for sphere systems are also readily verified using Lemma 9.40. Hence, one concludes that

$$S = s(\mathcal{K}) = \partial\Delta^{n+1},$$

$$S_e^0 := s(H_e^0),\ S_e^+ := s(H_e^+),\ S_e^- := s(H_e^-),\quad e \in E$$

actually gives rise to a sphere system. We have shown:

Theorem 9.41 *Given any simple OM, there exists a sphere system* $\mathcal{S} = (\partial\Delta^{n+1}, E, \mathcal{H})$ *such that* $\mathcal{O} = \mathcal{O}(\mathcal{S})$.

□

Note that we have shown, in fact, more than this. Our proof shows that all hyperspheres can be chosen to be "piecewise linear", i.e. to be the image of the hypershpere $\partial\Delta^n \subseteq \partial\Delta^{n+1}$ under a PLH $h : \partial\Delta^{n+1} \to \partial\Delta^{n+1}$.

9.5 Further Reading

As mentioned earlier, the "topological approach", characterizing oriented matroids in terms of sphere systems, is due to Folkman and Lawrence, cf.

J. FOLKMAN, J. LAWRENCE *Oriented Matroids*, Journal on Combinatorial Theory, Series B 25 (1978), pp. 199–236.

The stronger result, showing that in fact piecewise linear sphere systems suffice for representing OM-posets, is due to A. Mandel and J. Edmonds, cf.

A. MANDEL *Topology of Oriented Matroids*, Thesis, University of Waterloo, Canada (1981), supervised by J. Edmonds.

We refer to Mandel's work for a more detailed investigation of PL-sphere systems. For example, he proves that the axiom (iii) in the definition of sphere systems (Section 9.3) are redundant. The reader is warned, however, that our notation is slightly different from the one used by Mandel. This is due to the fact that we intended to work with the PL-terminology as used in Hudson's book

J.F.P. HUDSON *Piecewise Linear Topology*, W.A. Benjamin, Inc., New York (1969).

Independently of EDMOND's and MANDEL's work, the posets of ball complexes have been analysed with respect to shellability in

A. BJÖRNER *Posets, Regular CW-Complexes and Bruhat Order*, European Journal of Combinatorics 5 (1984), 7–16.

Bibliography

[1] M. ALFTER, W. KERN, A. WANKA (1988), *On Adjoints and Dual Matroids*, Journal on Combinatorial Theory, Series B, Vol 50 (2), 1990, 208–213.

[2] N. ALON (1986), *The Number of Polytopes, Configurations and Real Matroids*, Mathematica 33 (1986), 62–71.

[3] A. BACHEM (1983), *Convexity and Optimization in Discrete Structures*, in: P.M. Gruber, J.M. Wills (eds.), Convexity and Its Applications, Birkhäuser, Basel, 1983, 9–29.

[4] A. BACHEM, M. GRÖTSCHEL AND B. KORTE (1983), *Mathematical Programming, The State of the Art*, Springer, Heidelberg (1983).

[5] A. BACHEM, W. KERN (1984), *Convexity and Optimization in Discrete Structures*, in: R.W. Cottle, M.L. Kelmanson and B. Korte (eds.), Mathematical Programming (Proceedings of the International Congress on Mathematical Programming, Rio de Janeiro, 1981), North Holland, Amsterdam, 1984, 1–12.

[6] A. BACHEM, W. KERN (1985), *Adjoints of Oriented Matroids*, Combinatorica 6 (1985), 299–308.

[7] A. BACHEM, W. KERN (1986), *Extension Equivalence of Oriented Matroids*, European Journal of Combinatorics 7 (1986), 193–197.

[8] A. BACHEM, R. v. RANDOW (1978), *Integer Theorems of Farkas Lemma Type*, Methods of Operations Research 32 (1978), 19–28.

[9] A. BACHEM, A. WANKA (1985), *On Intersection Properties of (Oriented) Matroids (Extended Abstract)*, Methods of Operations Research 53 (1985), 227–229.

[10] A. BACHEM, A. WANKA (1986), *Euclidean Intersection Properties*, Journal of Combinatorial Theory B 47 (1989), 10–19.

[11] A. BACHEM, A. WANKA (1986), *Matroids without Adjoint*, Geometriae Dedicata 29 (1989), 311–315.

[12] A. BACHEM, A. WANKA (1988), *Separation Theorems for Oriented Matroids*, Discrete Mathematics 70 (1988), 303–310.

[13] C. BERGE (1962), *The Theory of Graphs and its Application*, J. Wiley, New York (1962).

[14] C. BERGE (1985), *Graphs*, North-Holland, Amsterdam (1985).

[15] C. BERGE, M. LAS VERGNAS (1984), *Transversals of Circuits and Acyclic Orientation in Graphs and Matroids*, Discrete Mathematics 50 (1984), 107–108.

[16] W. BIENA, R. CORDOVIL (1987), *An Axiomatic of Non-Radon Partitions of Oriented Matroids*, European Journal of Combinatorics 8 (1987), 1–4.

[17] N.L. BIGGS, E.K. LLOYD, R.J. WILSON (1976), *Graph Theory 1736–1936*, Oxford University Press, London, 1976.

[18] L.J. BILLERA, B.S. MUNSON (1984), *Polarity and Inner Products in Oriented Matroids*, European Journal of Combinatorics 5 (1984), 293–308.

[19] L.J. BILLERA, B.S. MUNSON (1984), *Triangulations of Oriented Matroids and Convex Polytopes*, SIAM Journal on Algebraic and Discrete Methods 5 (1984), 515–525.

[20] G. BIRKHOFF (1967), *Lattice Theory*, American Mathematical Society, Providence, Rhode Island, 1967.

[21] A. BJÖRNER (1984), *Posets, Regular CW–Complexes and Bruhat Order*, European Journal of Combinatorics 5 (1984), 7–16.

[22] A. BJÖRNER, P.H. EDELMAN, G.M. ZIEGLER (1990), *Hyperplane Arrangements with a Lattice of Regions*, Journal of Discrete and Computational Geometry, 5 (1990), 263–288.

[23] A. BJÖRNER, G.M. ZIEGLER (1991), *Combinatorial Stratification of Complex Arrangments*, Journal American Mathematical Society, 4 (1991).

[24] A. BJÖRNER, M. LAS VERGNAS, B. STURMFELS, N. WHITE, G. ZIEGLER (1991), *Oriented Matroids*, to appear in Encyclopedia of Mathematics, Cambridge University Press.

[25] R.G. BLAND (1974), *Complementary Orthogonal Subspaces of $\mathbb{R}^n$ and Orientability of Matroids*, Ph. D. Thesis, Cornell University, Ithaca, New York, 1974.

[26] R.G. BLAND (1977), *New Finite Pivoting Rules for the Simplex Method*, Mathematics of Operations Research 2 (1977), 103–107.

[27] R.G. BLAND (1977), *A Combinatorial Abstraction of Linear Programming*, Journal of Combinatorial Theory B 23 (1977), 33–57.

[28] R.G. BLAND, B.L. DIETRICH (1988), *On Abstract Duality*, Discrete Mathematics 70 (1988), 203–208.

[29] R.G. BLAND, D.L. JENSEN (1987), *Weakly Oriented Matroids*, Technical Report Ko732, School of OR/IE, Cornell University, Ithaca, New York, 1987.

[30] R.G. BLAND, M. LAS VERGNAS (1978), *Orientability of Matroids*, Journal of Combinatorial Theory B 24 (1978), 94–123.

[31] R.G. BLAND, M. LAS VERGNAS (1979), *Minty Colorings and Orientations of Matroids*, Annals of the New York Academy of Sciences 319 (1979), 86–92.

[32] J. BOKOWSKI, B. STURMFELS (1986), *Reell realisierbare orientierte Matroide*, Bayreuther Mathematische Schriften 21 (1986), 1–13.

[33] J. BOKOWSKI, B. STURMFELS (1986), *Coordinatization of Oriented Matroids*, Journal on Discrete and Computational Geometry 1 (1986), 293–306.

[34] J. BOKOWSKI, B. STURMFELS (1987), *Polytopal and Nonpolytopal Spheres — An Algorithmic Approach*, Israel Journal of Mathematics 57 (1987), 257–271.

[35] J. BOKOWSKI, B. STURMFELS (1989), *Computational Synthetic Geometry*, Lecture Notes in Mathematics, 1355, Springer - Verlag , Berlin - New York, 1989.

[36] B. BOLLOBÁS (1979), *Graph Theory*, Springer, Heidelberg, 1987.

[37] H. BRUGGESSER, P. MANI (1971), *Shellable Decompositions of Cells and Spheres*, **Mathematica Scandinavica 29 (1971), 197–205.**

[38] P.C.P. CARVALHO (1984), *On Certain Discrete Duality Models*, **Ph. D. Thesis, School of OR/IE, Cornell University, Ithaca, New York, 1984.**

[39] V. CHVÁTAL (1983), *Linear Programming*, **W.M. Freeman, New York (1983).**

[40] R. CORDOVIL (1980), *Sur les Orientations Acycliques des Géométries Orientés de Rang Trois*, **Annals of Discrete Mathematics 9 (1980), 243–246.**

[41] R. CORDOVIL (1982), *Sur un Théorèm de Séparation des Matroides Orientées de Rang Trois*, **Discrete Mathematics 40 (1982), 163–169.**

[42] R. CORDOVIL (1982), *Sur les Matroides Orientés de Rang Trois et les Arrangements de Pseudodroites dans le Plan Projectif Réel*, **European Journal of Combinatorics 3 (1982), 307–318.**

[43] R. CORDOVIL (1983), *Oriented Matroids and Geometric Sorting*, **Canadean Mathematical Bulletin 26 (1983), 351–354.**

[44] R. CORDOVIL (1983), *Oriented Matroids of Rank Three and Arrangements of Pseudolines*, **Annals of Discrete Mathematics 17 (1983), 219–223.**

[45] R. CORDOVIL (1985), *A Combinatorial Perspective on the Non-Radon Partitions*, **Journal of Combinatorial Theory A 38 (1985), 38–47; erratum, ibid. 40 (1985), 194.**

[46] R. CORDOVIL (1987), *Polarity and Point Extensions in Oriented Matroids*, **Linear Algebra and Its Applications 90 (1987), 15–31.**

[47] R. CORDOVIL, P. DUCHET (1986), *Séparation par une Droite dans les Matroides Orientées de Rang Trois*, **Discrete Mathematics 62 (1986), 103–104.**

[48] R. CORDOVIL, K. FUKUDA (1987), *Oriented Matroids and Combinatorial Manifolds*, **to appear in European Journal of Combinatorics.**

[49] R. CORDOVIL, A. GUEDES DE OLIVEIRA, M. LENOR MANEIRA (1988), *Parallel Projection of Matroid Spheres*, Portugaliae Mathematica 45 (1988), 1–10.

[50] R. CORDOVIL, M. LAS VERGNAS, A. MANDEL (1982), *Euler's Relation, Möbius Functions, and Matroid Identities*, Geometriae Dedicata 12 (1982), 147–162.

[51] R. CORDOVIL, I.P. DA SILVA (1985), *A Problem of McMullen on the Projective Equivalence of Polytopes*, European Journal of Combinatorics 6 (1985), 157–161.

[52] R. CORDOVIL, I.P. DA SILVA (1987), *Determining a Matroid Polytope by Non-Radon Partitions*, Linear Algebra and Its Applications 94 (1987), 55–60.

[53] H.H. CRAPO, G.-C. ROTA (1970), *On the Foundations of Combinatorial Theory: Combinatorial Geometries*, M.I.T. Press, London, 1970.

[54] D. CRYSTAL (1984), *Tag Systems: A Combinatorial Abstraction of Integral Dependence*, Working Paper, School of Operations Research and Industrial Engineering, Cornell University, New York, 1984.

[55] G. B. DANTZIG, *Reminiscenses about the Origin of Linear Programming*, in: A. Bachem, M. Grötschel and B. Korte: Mathematical Programming. The State of the Art, Springer (1983).

[56] G. B. DANTZIG, *Linear Programming and Extensions*, Princeton University Press, Princeton, N.J. (1963).

[57] G. DANARAJ, V. KLEE (1974), *Shellings of Spheres and Polytopes*, Duke Mathematical Journal 41 (1974), 443–451.

[58] A. DRESS (1983), *Matroids with Coefficients*, Working Paper, presented at the ZiF-Conference on Matroids, Bielefeld, 1983.

[59] A. DRESS (1986), *Duality Theory for Finite and Infinite Matroids with Coefficients*, Advances in Mathematics 59 (1986), 97–123.

[60] A. DRESS (1986), *Chirotopes and Oriented Matroids*, Bayreuther Mathematische Schriften 21 (1986), 14–68.

[61] A. DRESS, W. WENZEL (1987), *Geometric Algebra for Combinatorial Geometries*, to appear in Advances in Mathematics.

[62] A. DRESS, W. WENZEL (1988), *Endliche Matroide mit Koeffizienten*, Bayreuthische Mathematische Schriften 26 (1988), 37–98.

[63] A. DRESS, W. WENZEL (1988), *Grassmann-Plücker Relations and Matroids with Coefficients*, to appear in Advances in Mathematics.

[64] DUGUNDJI (1966), *Topology*, Allyn and Bacon Inc., Boston, 1966.

[65] P.H. EDELMAN (1982), *The Lattice of Convex Sets of an Oriented Matroid*, Journal of Combinatorial Theory B 33 (1982), 239–244.

[66] P.H. EDELMAN (1984), *A Partial Order on the Regions of $\mathbb{R}^n$ dissected by Hyperplanes*, Transactions of the American Mathematical Society 283 (1984), 617–631.

[67] P.H. EDELMAN (1984), *The Acyclic Sets of an Oriented Matroid*, Journal of Combinatorial Theory B 36 (1984), 26–31.

[68] P.H. EDELMAN, R. JAMISON (1985), *The Theory of Convex Geometries*, Geometriae Dedicata 19 (1985), 247–270.

[69] P.H. EDELMAN, J.N. WALKER (1985), *The Homotopy Type of Hyperplane Points*, Proceedings of the American Mathematical Society 94 (1985), 329–332.

[70] U. FAIGLE (1985), *Orthogonal Sets, Matroids, and Theorems of the Alternative*, Bulletin U.M.I. 4–B (1985), 139–153.

[71] J. FARKAS (1902), *Theorie der einfachen Ungleichungen*, Journal für Reine und Angewandte Mathematik 124 (1902), pp. 1–27.

[72] W.E. FENTON (1982), *Axiomatic of Convexity Theory*, Ph. D. Thesis, Purdue University, Indiana, 1982.

[73] W.E. FENTON (1987), *Completeness in Oriented Matroids*, Discrete Mathematics 66 (1987), 79–90.

[74] R. FLETCHER (1969), *Optimization*, Academic Press (1969).

[75] J. FOLKMAN, J. LAWRENCE (1978), *Oriented Matroids*, Journal of Combinatorial Theory B 25 (1978), 199–236.

[76] J.-B. FOURIER (1936), *Solution d'une Question Particulière du Calcul des Inégalités*, Oevres II (1936), pp. 317–328.

[77] K. FUKUDA (1982), *Oriented Matroid Programming*, Ph. D. Thesis, University of Waterloo, Waterloo, Ontario, 1982.

[78] K. FUKUDA, A. TAMURA (1986), *Dualities in Signed Vector Systems*, to appear in Portugaliae Mathematica.

[79] K. FUKUDA, A. TAMURA (1988), *Local Deformation and Orientation Transformation in Oriented Matroids*, Ars Combinatorica A 25 (1988), 243–258.

[80] K. FUKUDA, A. TAMURA (1989), *Characterizations of *-Families*, Journal of Combinatorial Theory B, 47 (1989), no. 1, 107–110..

[81] K. GÖDEL (1931), *On Formally Undecidable Propositions of Principia Mathematica and Related Systems I*, Monatshefte für Mathematik und Physik 38 (1931), pp. 173–198, reprinted in S. Ferman et al, "Kurt Gödel: Collected Works I: Publications 1929–1936", Oxford University Press (1986).

[82] J.E. GOODMAN, R. POLLACK (1980), *Proof of Grünbaum's Conjecture on the Stretchability of Certain Arrangements of Pseudolines*, Journal of Combinatorial Theory A 29 (1980), 385–390.

[83] J.E. GOODMAN, R. POLLACK (1981), *Three Points do not determine a (Pseudo-) Plane*, Journal of Combinatorial Theory A (1981), 215–218.

[84] J.E. GOODMAN, R. POLLACK (1982), *A Theorem of Ordered Duality*, Geometriae Dedicata 12 (1982), 63–74.

[85] J.E. GOODMAN, R. POLLACK (1983), *Multidimensional Sorting*, SIAM Journal on Computing 12 (1983), 484–503.

[86] J.E. GOODMAN, R. POLLACK (1984), *Semispaces of Configurations, Cell Complexes of Arrangements*, Journal of Combinatorial Theory A 37 (1984), 257–293.

[87] J.E. GOODMAN, R. POLLACK (1985), *A Combinatorial Version of the Isotopy Conjecture*, Discrete Geometry and Convexity (New York, 1982), 12–19, Annals of the New York Acadamy of Sciences 440, 1985.

[88] J.E. GOODMAN, R. POLLACK (1985), *Polynomial Realization of Pseudoline Arrangements*, Communications in Pure and Applied Mathematics 38 (1985), 725–732.

[89] J.E. GOODMAN, R. POLLACK (1986), *Upper Bounds for Configurations and Polytopes in $\mathbb{R}^d$*, Discrete and Computational Geometry 1 (1986), 219–227.

[90] P. GOOSSENS (1987), *Combinatorial and Homological Properties of some Partially Ordered Sets Arising in Geometry*, Ph.D.Thesis, Liege University, Liege, 1987.

[91] M. GOUDRAN, M. MINOUX, S. VAJDA (1984), *Graphs and Algorithms*, J. Wiley, New York, 1984.

[92] C. GRÄTZER (1978), *General Lattice Theory*, Birkhäuser, Basel, 1978.

[93] C. GREENE (1977), *Acyclic Orientations (Notes)* in: M. Aigner (ed.), Higher Combinatorics, Reidel, Dordrecht, 1977, 65–68.

[94] C. GREENE, T.L. MAGNANTI (1974), *Some Abstract Pivot Algorithms*, Department of Mathematics, M.I.T., Cambridge, 1974.

[95] C. GREENE, T. ZASLAVSKY (1983), *On the Interpretation of Whitney Numbers through Arrangements of Hyperplanes, Zonotopes, Non-Radon Partitions, and Orientations of Graphs*, Transactions of the American Mathematical Society 280 (1983), 97–126.

[96] B. GRÜNBAUM (1967), *Convex Polytopes*, Wiley, London, 1967.

[97] B. GRÜNBAUM (1972), *Arrangements and Spreads*, Regional Conference Series in Mathematics 10, American Mathematical Society 114, Providence, Rhode Island, 1972.

[98] L. GUTIERREZ NOVOA (1965), *On n-ordered Sets and Order Completeness*, Pacific Journal of Mathematics 15 (1965), 1337–1345.

[99] Y.O. HAMIDOUNE, M. LAS VERGNAS (1984), *Jeux de Commutation Orientées sur les Graphes et les Matroides*, C.R. Academic Science Paris 298 (1984), 497–499.

[100] Y.O. HAMIDOUNE, M. LAS VERGNAS (1986), *Directed Switching Games on Graphs and Matroids*, Journal of Combinatorial Theory B 40 (1986), 237–269.

[101] K. HANDA (1990), *A Characterization of Oriented Matroids in Terms of their Topes*, European Journal of Combinatorics 11, (1990), 41–46.

[102] W. HOCHSTÄTTLER (1990), *Shellability of Oriented Matroids* Proceedings of ICPO, University of Waterloo, Canada (1990), 275–281.

[103] J.F.P. HUDSON (1969), *Piecewise Linear Topology*, W.A. Benjamin, Inc., New York, 1969.

[104] B. JAGGI, P. MANI-LEVITSKA, B. STURMFELS, N.L. WHITE, (1989), *Uniform Oriented Matroids without the isotopy property*, Journal of Discrete andComputational Geometry 4 (1989), no. 2,97–100.

[105] D.L. JENSEN (1985), *Coloring and Duality: Combinatorial Augmentation Methods*, Ph. D. Thesis, Cornell University, Ithaca, New York, 1985.

[106] J.P. JONES and MATIJASEVIC (1984), *Register Machine Proof of the Theorem on Exponential Diophantine Representation of Enumerable Sets*, Journal on Symbolic Logic 49 (1984), pp. 818–829,.

[107] J. KAHN (1987), *On Lattices with Möbius Function* $\pm 1, 0$, Descrete and Computational Geometry 2 (1987), 1–8.

[108] W. KERN (1985), *Verbandstheoretische Dualität in kombinatorischen Geometrien und orientierten Matroiden*, Ph.D.Thesis, Mathematisches Institut, Universität zu Köln, 1985.

[109] E. KLAFSZKY, T. TERLAKY, (1989), *Some Generalizations of the Crisscross Method for the Linear Complementarity Problem of Oriented Matroids*, Combinatorica 9 (2) (1989), 189–198.

[110] E. KLAFSZKY, T. TERLAKY, (1987), *Remarks on the Feasibility Problem of Oriented Matroids*, Annales Universitatis Scientientiarium Budapestinensis de Rolando Eöfvös Nominatae, Sectio Computatorica, Tomus VII, 1987, 155–157.

[111] M. LAS VERGNAS (1975), *Matroides Orientables*, C.R. Academic Science Paris 280 (1975), 61–64.

[112] M. LAS VERGNAS (1977), *Acyclic and Totally Cyclic Orientations of Combinatorial Geometries*, Discrete Mathematics 20 (1977), 51–61.

[113] M. LAS VERGNAS (1978), *Extensions Ponctuelles d'une Géometrie Combinatoire*, in: Problèmes Combinatoires et Théorie

des Graphes, Actes du Colloque International C.N.R.S. 260, Orsay, 1976, Paris, 1978, 263–268.

[114] M. LAS VERGNAS (1978), *Sur les Activités des Orientations d'une Géométrie Combinatoire*, in: Codes et Hypergraphes (Actes du Colloque Math. Discrètes, Bruxelles, 1978), Cahiers du Centre de Recherche Opérationelle 20, Bruxelles, 1978, 293–300.

[115] M. LAS VERGNAS (1978), *Bases in Oriented Matroids*, Journal of Combinatorial Theory B 25 (1978), 283–289.

[116] M. LAS VERGNAS (1980), *Convexity in Oriented Matroids*, Journal of Combinatorial Theory B 29 (1980), 231–243.

[117] M. LAS VERGNAS (1981), *Oriented Matroids as Signed Geometries Real in Corank 2*, Actes du Colloque d'Eger (Hongrie), 1981.

[118] M. LAS VERGNAS (1984), *A Correspondance between Spanning Trees and Orientations in Graphs*, in: B. Bollobas (ed.), Graph Theory and Combinatorics (Proceedings of the Cambridge Combinatorial Conference in Honour of Paul Erdös, Cambridge, 1983), Academic Press, London, 1984, 233–238.

[119] M. LAS VERGNAS (1986), *Order Properties of Lines in the Plane and a Conjecture of Ringel*, Journal of Combinatorial Theory B 41 (1986), 246–249.

[120] J. LAWRENCE (1975), *Oriented Matroids*, Ph. D. Thesis, Washington University, Seattle, Washington, 1975.

[121] J. LAWRENCE (1982), *Oriented Matroids and Multiply Ordered Sets*, Linear Algebra and Its Applications 48 (1982), 1–12.

[122] J. LAWRENCE (1983), *Lopsided Sets and Orthant-intersection by Convex Sets*, Pacific Journal of Mathematics 104 (1983), 155–173.

[123] J. LAWRENCE (1984), *Shellability of Oriented Matroid Complexes.* working paper, George Mason University, Fairfax, Virginia.

[124] J. LAWRENCE, L. WEINBERG (1981), *Unions of Oriented Matroids*, Linear Algebra and Its Applications 41 (1981), 183–200.

[125] D. LUENBERGER (1973), *Introduction to Linear and Nonlinear Programming*, Addison Wesley (1973).

[126] A. MANDEL (1982), *Topology of Oriented Matroids*, Ph. D. Thesis, University of Waterloo, Waterloo, Ontario, 1982.

[127] G. MCCORMICK (1983), *Nonlinear Programming*, John Wiley (1983).

[128] P. MCMULLEN (1979), *Transforms, Diagrams, and Representations*, in: J. Tölke, J.M. Wills (eds.), Contribution to Geometry (Proceedings of Geometry Symposium, Siegen, 1978), Birkhäuser, Basel, 1979, 92–130.

[129] P. MCMULLEN, J.C. SHEPHARD (1971), *Convex Polytopes and the Upper Bound Conjecture*, Cambridge University Press, Cambridge, 1971.

[130] P.S. MEI (1971), *Axiomatic Theory of Linear and Convex Closure*, Ph. D. Thesis, Purdue University, Indiana, 1971.

[131] D.A. MILLER (1983), *A Class of Topological Oriented Matroids with some Applications to Nonlinear Programming*, Ph. D. Thesis, Berkeley University, California, 1983.

[132] D.A. MILLER (1987), *Oriented Matroids from Smooth Manifolds*, Journal of Combinatorial Theory B 43 (1987), 173–186.

[133] G.J. MINTY (1966), *On the Abstract Foundations of the Theory of Directed Linear Graphs, Electrical Networks and Network Programming*, Journal of Mathematical Mech. 15 (1966), 485–520.

[134] W.D. MORRIS, M.J. TODD (1988), *Symmetry and Positive Definiteness in Oriented Matroids*, European Journal of Combinatorics 9 (1988), 121–130.

[135] B.S. MUNSON (1981), *Face Lattices of Oriented Matroids*, Ph. D. Thesis, Cornell University, Ithaca, New York, 1981.

[136] K. MURTY (1976), *Linear and Combinatorial Programming*, John Wiley (1976).

[137] C.H. PAPADIMITRIOU, K. STEGLITZ (1982), *Combinatorial Optimization*, Prentice Hall, Englewood Cliffs, N.J. 1982.

[138] R.T. ROCKAFELLAR (1969), *The Elementary Vectors of a Subspace of $\mathbb{R}^n$*, in: R.C. Bose and T.A. Dowlings (eds.), Combinatorial Theory and Its Applications (Proceedings of the Chapel Hill Conference), North Carolina University Press, Chapel Hill, North Carolina, 1969, 104–127.

[139] J.P. ROUDNEFF (1988), *Reconstruction of the Orientation Class of an Oriented Matroid.* **European Journal of Combinatorics 9 (1988), no.5, 423–429.**

[140] J.P. ROUDNEFF (1989), *Inseparability Graphs of Oriented Matroids.* **Combinatorica 9 (1) (1989), 75–84.**

[141] J.P. ROUDNEFF (1986), *On the Number of Triangles in Simple Arrangements of Pseudolines in the Real Projective Plane,* **Discrete Mathematics 60 (1986), 243–251.**

[142] J.P. ROUDNEFF (1987), *Matroides Orientées et Arrangements de Pseudodroites,* **Ph. D. Thesis, Paris, 1987.**

[143] J.P. ROUDNEFF (1988), *Tverberg-type Theorems for Pseudoconfigurations of Points in the Plane,* **European Journal of Combinatorics 9 (1988), 189–198.**

[144] J.P. ROUDNEFF, B. STURMFELS, *Simplical Cells in Arrangements and Mutations of Oriented Matroids,* **Geometriae Dedicata 27 (1988), 153–170.**

[145] M.E. RUDIN (1958), *An Unshellable Triangulation of a Tetrahedron,* **Bull. Amer. Math. Soc. 64 (1958), 90–91.**

[146] A. SCHRIJVER (1986), *Theory of Linear and Integer Programming,* **John Wiley, New York (1986).**

[147] I. SHEMER (1982), *Neighbourly Polytopes,* **Israel Journal of Mathematics 43 (1982), 291–314.**

[148] I.P. DA SILVA (1987), *Quelques Propriétés des Matroides Orientés,* **Ph. D. Thesis, Paris, 1987.**

[149] J. STOER, CHR. WITZGALL (1970), *Convexity and Optimization in Finite Dimension I,* **Springer, Berlin, 1970.**

[150] B. STURMFELS (1986), *Central and Parallel Projections of Polytopes,* **Discrete Mathematics 62 (1986), 315–318.**

[151] B. STURMFELS (1987), *Computational Synthetic Geometry,* **Ph.D.Thesis, University of Washington, Seattle, 1987.**

[152] B. STURMFELS (1987), *Oriented Matroids and Combinatorial Convex Geometry,* **Ph. D. Thesis, Universität Darmstadt, Darmstadt, 1987.**

[153] B. STURMFELS (1987), *Boundary Complexes of Convex Polytopes cannot be Characterized Locally*, Journ. London Math. Soc. 35 (1987), 257–269.

[154] B. STURMFELS (1987), *Cyclic Polytopes and d-Order Curves*, Geometriae Dedicata 24 (1987), 103–107.

[155] B. STURMFELS (1987), *On the Decidablity of Diophantine Problems in Combinatorial Geometry*, Bulletin of the American Mathematical Society 17 (1987), 121–124.

[156] B. STURMFELS (1988), *Some Applications of Affine Gale Diagrams to Polytopes with few Vertices*, SIAM Journal of Discrete Mathematics 1 (1988), 121–133.

[157] B. STURMFELS (1988), *Totally Positive Matrices and Cyclic Polytopes*, Linear Algebra and Its Applications 107 (1988), 275–281.

[158] B. STURMFELS (1988), *Neighbourly Polytopes and Oriented Matroids*, Preprint, European Journal of Combinatorics 9 (1988), no.6, 537–546.

[159] T. TERLAKY (1987), *A Finite Crisscross Method for Oriented Matroids*, Journal of Combinatorial Theory B 42 (1987), 319–327.

[160] M. TODD (1984), *Complementarity in Oriented Matroids*, SIAM Journal on Algebraic and Discrete Methods 5 (1984), 467–485.

[161] M. TODD (1985), *Linear and Quadratic Programming in Oriented Matroids*, Journal of Combinatorial Theory B 39 (1985), 105–133.

[162] A.M. TURING (1937), *On Computable Numbers with an Application to the Entscheidungsproblem*, Proceedings London Mathematical Society 42 (1937), pp. 230–265, reprinted in M. Davis, "The Undecidable-Basic Papers on Undecidable Propositions, Unsolvable Problems and Computable Functions", Hewlett: Raven Press (1965)..

[163] W.T. TUTTE (1984), *Graph Theory*, Encylopedia of Mathematics and its Applications, Vol. 21, Addison Wesley, Reading, Mass. (1984)

[164] M. WAGOWSKI (1989), *Matroid Signatures Coordinatizable over a Semiring*, European Journal of Combinatorics 10 (4) (1989), 393–398.

[165] ZH. WANG (1987), *A Finite Conformal-elimination-free Algorithm for Oriented Matroid Programming*, Chinese Annals of Mathematics, 8B(1) (1987)

[166] A. WANKA (1986), *Matroiderweiterungen zur Existenz endlicher LP–Algorithmen, von Hahn–Banach Sätzen und Polarität von orientierten Matroiden*, Ph.D.Thesis, Mathematisches Institut, Universität zu Köln, Köln, 1986.

[167] D.J.A. WELSH (1976), *Matroid Theory*, Academic Press, London, 1976.

[168] W. WENZEL (1988), *A Group-Theoretic Interpretation of Tutte's Homotopy-Theory*, to appear in Advances in Mathematics.

[169] N. WHITE (ED.) (1986), *Theory of Matroids*, Encyclopedia of Mathematics and Its Applications, Cambridge University Press, Cambridge, 1986.

[170] N. WHITE (1989), *A Non-Uniform Oriented Matroid which violates the Isotopy Property*, Discrete Computational Geometry 4 (1989), 1–2.

[171] H. WHITNEY (1935), *On the Abstract Properties of Linear Dependence*, American Journal of Mathematics 57 (1935), 509–533.

[172] T. ZASLAVSKY (1975), *Facing up to Arrangements: Face-Count Formulas for Partitions of Space by Hyperplanes*, Memoires of the American Mathematical Society 154 (1975).

[173] T. ZASLAVSKY (1985), *Extremal Arrangements of Hyperplanes*, Discrete Geometry and Convexity (New York, 1982), 69–87, Annals of the New York Acadamy of Sciences 440, 1985.

[174] G.M. ZIEGLER (1988), *The Face Lattice of Hyperplane Arrangements*, Discrete Mathematics 74 (1988), 233–238.

[175] G.M. ZIEGLER (1991), *Some Minimal Non-orientable Matroids of Rank 3*, Geometriae Dedicata 38 (1991), 365–371.

Symbols and Expressions

Index

Universitext

Aksoy, A., Khamsi, M.A: Methods in Fixed Point Theory
Aupetit, B.: A Primer on Spectral Theory
Berger, M.: Geometry I
Berger, M.: Geometry II
Bliedtner, J. / Hansen, W.: Potential Theory
Booss, B. / Bleecker, D.D.: Topology and Analysis
Cecil, T.E.: Lie Sphere Geometry
Chandrasekharan, K.: Classical Fourier Transforms
Charlap, L.S.: Bieberbach Groups and Flat Manifolds
Chern, S.: Complex Manifolds without Potential Theory
Cohn, H.: A Classical Invitation to Algebraic Numbers and Class Fields
Curtis, M.L.:Abstract Linear Algebra
Curtis, M.L.: Matrix Groups
Dalen, D. van: Logic and Structure
Devlin, K.J.: Fundamentals of Contemporary Set Theory
Dimca, A.: Singularities and Topology of Hypersurfaces
Edwards, R.E.: A Formal Background to Higher Mathematics I a, and I b
Edwards, R.E.: A Formal Background to Higher Mathematics II a, and II b
Emery, M.: Stochastic Calculus in Manifolds
Foulds, L.R.: Graph Theory and Applications
Fuks, D.B. / Rokhlin, V.A.: Beginner's Course in Topology
Frauenthal, J.C.: Mathematical Modeling in Epidemiology
Gallot, S. / Hulin, D. / Lafontaine, J.: Riemannian Geometry
Gardiner, C.F.: A First Course in Group Theory
Garding, L. / Tambour, T.: Algebra for Computer Science
Godbillon, C: Dynamical Systems on Surfaces
Goldblatt, R.: Orthogonality and Spacetime Geometry
Hájek, P. / Havránek, T.: Mechanizing Hypothesis Formation
Hlawka, E. / Schoißengeier, J. / Taschner, R.: Geometric and Analytic Number Theory
Humi, M. / Miller, W.: Second Course in Ordinary Differential Equations for Scientists and Engineers
Hurwitz, A. / Kritikos, N.: Lectures on Number Theory
Iversen, B.: Cohomology of Sheaves
Jones, A./Morris, S.A./Pearson, K.: Abstract Algebra and Famous Impossibilities
Kelly, P. / Matthews, G.: The Non-euclidean Hyperbolic Plane
Kempf,G: Complex Abelian Varieties and Theta Functions
Kostrikin, A.I.: Introduction to Algebra
Krasnoselskii, M.A. / Pokrovskii, A.V.: Systems with Hysteresis
Luecking, D.H. / Rubel, L.A.: Complex Analysis. A Functional Analysis Approach
Mac Lane, S./Moerdijk I.: Sheaves in Geometry an Logic
McCarthy, P.J.: Introduction to Arithmetical Functions
Marcus, D.A.: Number Fields
Meyer, R.M.: Essential Mathematics for Applied Fields
Meyer-Nieberg, P.: Banach Lattices
Mines, R. / Richman, F./ Ruitenberg, W.: A Course in Constructive Algebra

Moise, E.E.: Introductory Problem Courses in Analysis and Topology
Montesinos, J.M.: Classical Tessellations and Three-Manifolds
Nikulin, V.V. / Shafarevich, I.R.: Geometries and Groups
Oksendal, B.: Stochastic Differential Equations
Porter, J.R. / Woods, R.G.: Extensions and Absolutes of Hausdorff Spaces
Rees, E.G.: Notes on Geometry
Reisel, R.B.: Elementary Theory of Metric Spaces
Rey, W.J.J.: Introduction to Robust and Quasi-Robust Statistical Methods
Rotman, J.J.: Galois Theory
Rybakowski, K.P.: The Homotopy Index and Partial Differential Equations
Samelson, H.: Notes on Lie Algebras
Smith, K.T.: Power Series from a Computational Point of View
Smorynski, C.: Self-Reference and Modal Logic
Smorynski, C.: Logical Number Theory I: An Introduction
Stroock, D.W.: An Introduction to the Theory of Large Deviations
Sunada, T.: The Fundamental Group and the Laplacian (to appear)
Sunder, V.S.: An Invitation to von Neumann Algebras
Tondeur, P.: Foliations on Riemannian Manifolds
Verhulst, F.: Nonlinear Differential Equations and Dynamical Systems
Zaanen, A.C.: Continuity, Integration and Fourier Theory

Printing and binding: Druckhaus Beltz, Hemsbach